THAI SILK

JENNIFER SHARPLES

Published by Post Books,
a division of The Post Publishing Public Co. Ltd.
136 Na Ranong Road, off Sunthorn Kosa Road,
Klong Toey, Bangkok 10110, Thailand.

First Edition 1994
Copyright ©1993 Jennifer Sharples
ISBN: 974-202-010-8
Editors: David Pratt, Catherine Purananda
Designer: Norman Bright
Set in Garamond
Printed by Allied Printers, Bangkok

All rights reserved. No part of this publication may be reproduced, stored in a retrieval system, or transmitted, in any form or by any means, electronic, mechanical, photocopying, recording or otherwise, without prior permission of the publisher.

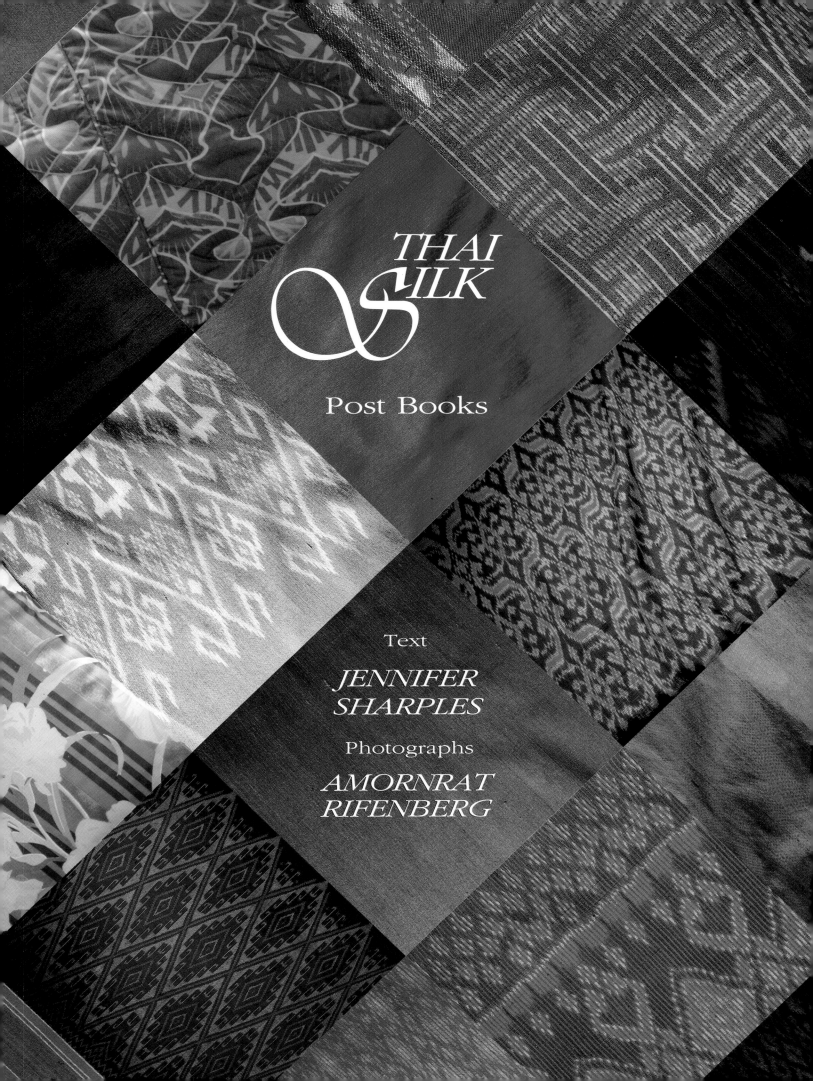

THAI SILK

Post Books

Text

JENNIFER SHARPLES

Photographs

AMORNRAT RIFENBERG

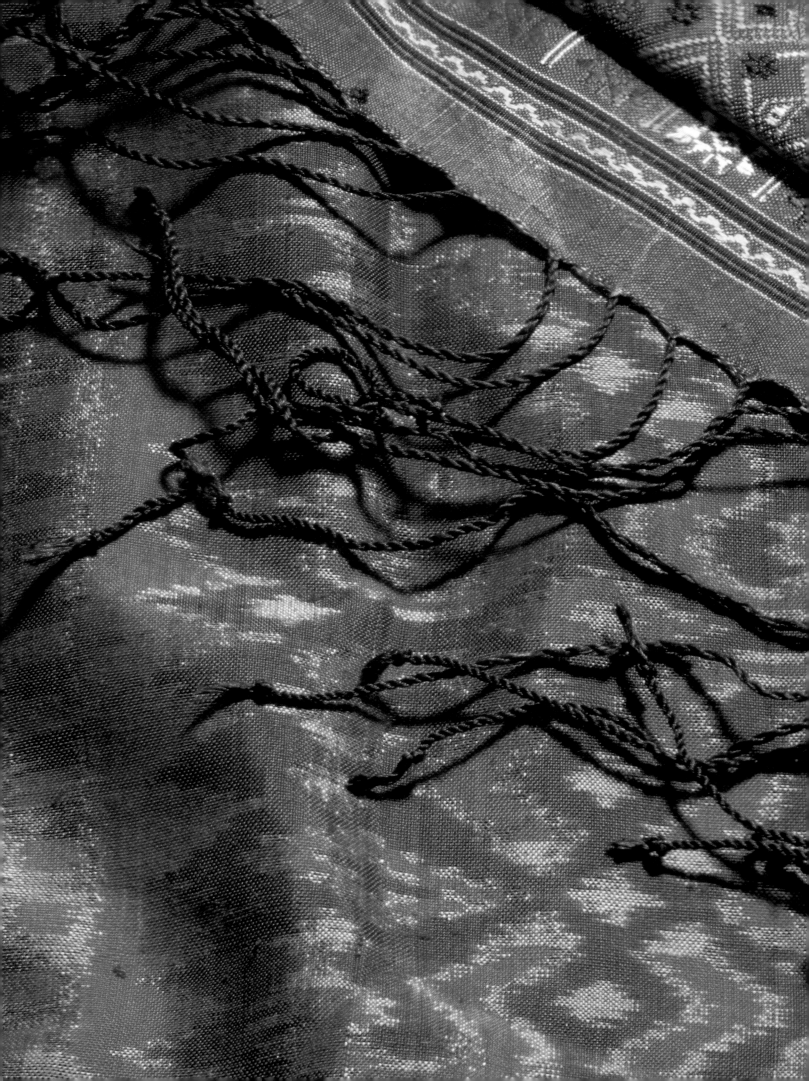

Weaving Up History

PAGES 10-25

From Cocoon To Loom

PAGES 26-49

On Royal Command

PAGES 50-65

Magnificent Mudmee

PAGES 66-81

The Fashion World Discovers Thai Silk

PAGES 82-101

Sumptuous Decors Of Silk

PAGES 102-117

Collecting Silk Textiles

PAGES 118-141

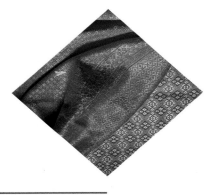

The World Of Thai Silk

PAGES 142-155

GLOSSARY OF TERMS .... PAGES 156-157

SELECTED BIBLIOGRAPHY .... PAGES 158

ACKNOWLEDGEMENTS .... PAGES 160

Left: *Silk weaving is one of Thailand's oldest surviving handicrafts. This contemporary mural painting depicts an old style rural scene where villagers in traditional costumes are occupied with reeling, dyeing and weaving silk.*

Right: *A 19th century print depicts a youthful King Chulalongkorn (King Rama V, 1868-1910) attired in his coronation robes, which were made from the finest and richest silks.*

Weaving Up History

Acclaimed the "Queen of Fibres", silk boasts a fascinating and romantic heritage, its origins embroidered in a wealth of history and legends. Evidence, entangled with legend, reveals silk to have been a form of currency, a measure of wealth, a medium of art, and a trade item largely responsible for bringing East and West into contact.

An ancient Chinese legend recounts how tens of centuries ago, the Great Yellow Emperor Huang Ti (2640 B.C.), China's first imperial ruler, was celebrating a victory over his enemies. The silkworm goddess descended to earth and presented him with a bundle of silvery white silk thread, which was then transformed into a beautiful fabric.

A conflicting legend attributes the discovery of silk to Hsi Ling Shih, the consort of Emperor Huang Ti. The story goes that as Hsi Ling Shih sat languidly sipping her tea beneath the mulberry trees in the Imperial Gardens,

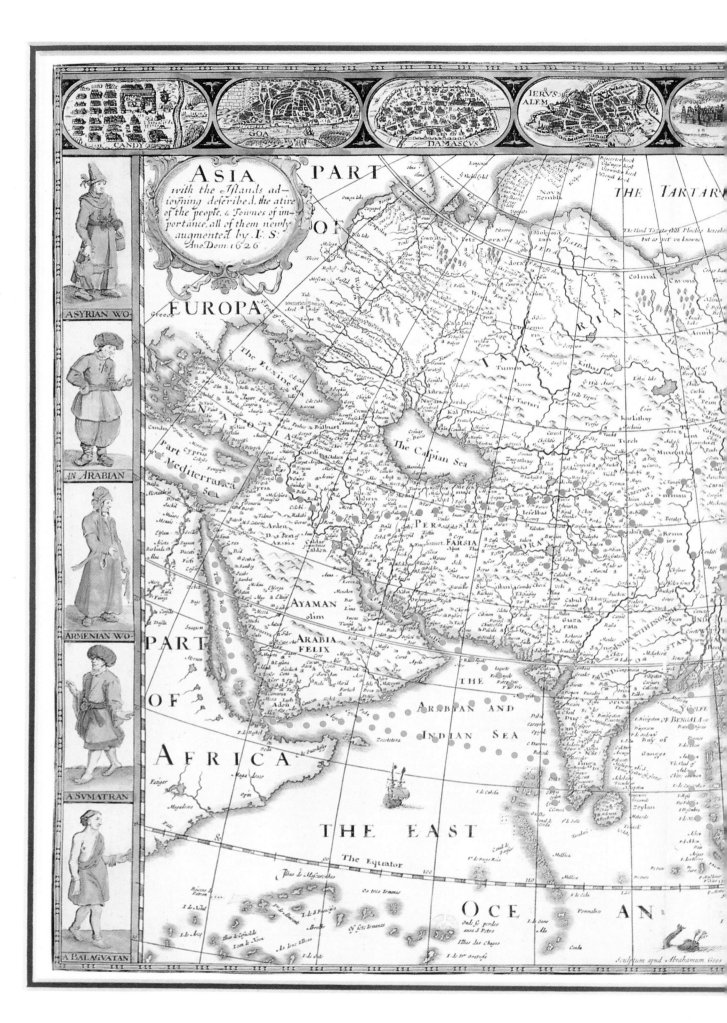

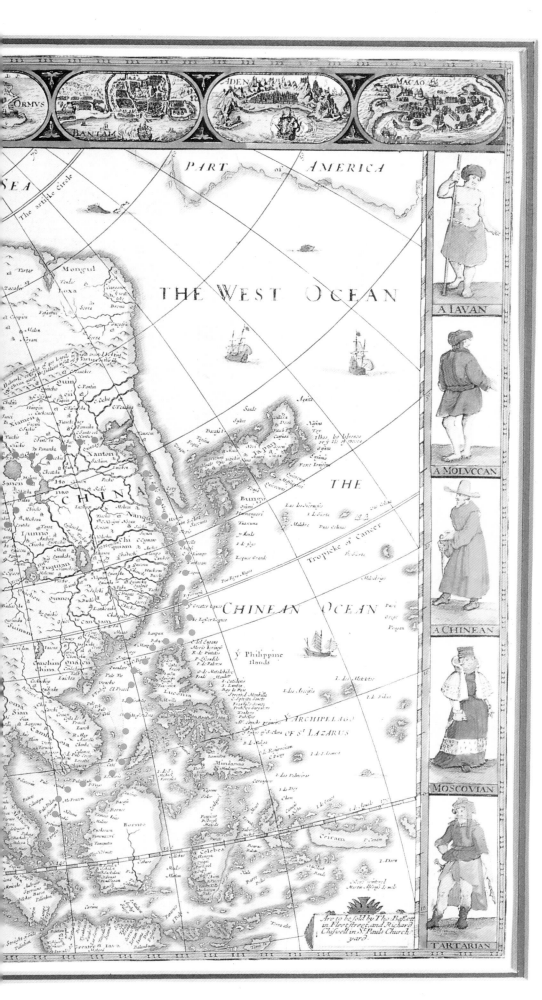

The Silk Road, which opened up in the second century B.C., was the strand that linked the ancient world's two superpowers, China and Rome. It was not so much a road as a moving network of caravan trails between remote trading posts. Although many other commodities were traded, silk was the most precious, giving the road its illustrious name. This 15th century map traces the main overland routes and the early sea routes.

········· sea routes

········· overland routes

a creamy coloured cocoon landed in her cup. To her astonishment, it unravelled to form a long and delicate thread. Delighted with its beauty, she gathered thread from thousands of other cocoons and wove them into a ceremonial robe for the Emperor.

Evidence indicates that sericulture and silk weaving techniques originated in China four to five thousand years ago. In Shangtung province, the Imperial Court established factories to weave silk fabrics both for ceremonial use and as gifts for foreign powers. Chinese emperors profited handsomely from taxes levied on silk, and the fabric became so treasured that it was used as a measure of currency and reward. Lifelike carved jade silkworms have been discovered amongst Shang Dynasty (1523-1027 B.C.) relics, hieroglyphs of silkworms and silk appear on oracle inscriptions of the same period, and actual silk remnants were found in a tomb dating to the Warring States period (475-221 B.C.).

THE THREAD UNRAVELS

For hundreds of years, the Chinese jealously guarded the secret art of sericulture; death by torture was the penalty for those who disclosed it. As a result, sericulture spread slowly. It reached Khotan on the Afghanistan border around 140 B.C., when an Imperial princess married a prince of Khotan and smuggled out silkworm eggs by hiding them in her headdress. The eggs were part of her considerable dowry. According to legend, the art spread east to Japan around A.D. 195, but under less romantic conditions. Two Chinese concubines were kidnapped and coerced into revealing the secret, which was then exchanged for a large reward. Western legend claims that in A.D. 550, the Roman Emperor Justinian persuaded two monks to journey to China to search for the secret. Two years later they returned; one monk carried white mulberry seeds in a pouch, and the other concealed the silkworm eggs in his bamboo cane. It is claimed that the first European silkworms descended from those very eggs.

THE BEDAZZLED WEST

Historical references to silk, the Far Eastern wonder, appear through the centuries. Two-thousand-year-old Hindu epics refer to the precious fabric, both Isaiah and Ezekiel make mention of it in the Old Testament, and in the fourth century B.C. Aristotle describes the silk caterpillar as a "horned

Above: *An 18th century print shows a royal prince of Siam, attired in a gorgeous silk costume and flanked by his mother and a page, based on the impressions of a French visitor to the Kingdom of Siam.*

worm". The Greeks began producing silk after Alexander the Great (356-323 B.C.) conquered Persia; and in 330 B.C. China was referred to as the "Land of Silk" by the Egyptians. Such was its value in the Roman Empire that during Julius Caesar's reign (61-44 B.C.) it sold for its weight in gold. Emperor Tiberius passed laws to curb lavish displays of the material and prohibited men from wearing silk on the ground that it was effeminate.

Nevertheless, silk was held to be a sacred commodity in the West until the Han Dynasty (205 B.C. to A.D. 220), when the perilous Silk Road came into existence. It began in Xi'an in Shaanxi Province and ran across the Central Asian continent, through Persia to the Mediterranean Sea. From there, the silk reached other destinations by ship. Few completed the whole journey; caravan loads were passed from trader to trader and often exchanged for commodities such as gold, jade, wool, horses and glass. The Silk Road linked the East and the West where, each equally enigmatic to the other, the Chinese and Roman empires reigned at either end of the route.

Around 1275, the Venetian explorer Marco Polo returned to Italy after years of adventure in China and Asia, bringing with him glorious silks and sericulture knowledge. From the 14th century, silk production thrived in Europe, and both Italy and France became important silk centres. In 1493, the Duchess of Milan wrote to a friend that she was having dresses embroidered in silk with patterns that had been designed for frescoes by Leonardo da Vinci. By the 17th century, sericulture had reached the American Colonies with the early settlers and thereby had encircled the globe.

SILK IN SIAM

Archaeological discoveries in the village of Ban Chiang in the northeast province of Udon Thani have led experts to believe that Thailand's sericulture history may be as old as China's. An extensive burial site yielded evidence of a complex civilisation dating back over 4,000 years. Ban Chiang people cultivated crops and produced ornaments and bronze tools. Excavations also revealed a cluster of unwoven and undyed silk thread. Similar silk thread remnants were found in the pre-historic area of Ban Nadi in Nong Han, Udon Thani. Both discoveries strongly suggest that sericulture existed amongst Thailand's prehistoric civilisations.

Between the fifth and seventh centuries, the importance of the Silk

Pages 16-17: *Camel caravans, merchants, monks, pilgrims, entertainers and spies travelled the Silk Road together with well dressed emissaries on horseback, such as these Chinese horsemen.*

Above: *A royal page of Siam is dressed in a casual costume of the late 18th century in silk garments as befitting a member of the Royal Household.*

Right: *Queen Saovabha, consort of King Chulalongkorn (Rama V, 1868-1910). She is dressed in the style that was favoured by aristocratic ladies of the late 19th and early 20th centuries — a combination of a Western-style lace blouse and a Thai-style silk* pha nung.

Pages 20-21: *The story of silk began in China thousands of years ago. This ancient painting on silk shows ladies of the Chinese Imperial Court unrolling lengths of the country's most precious commodity, silk.*

Road began to decline as sea routes were discovered that proved less hazardous, promoting trade between China, Southeast Asia, India and the West. Early trade along the silk sea route passed through Southeast Asia where Chinese silks and porcelains were highly desirable commodities. Archaeological evidence suggests that sea traders had reached the early Mon settlers of the Dvaravati Kingdom in Siam who controlled settlements along the river plains and the Gulf of Siam. It is highly probable that silk would have been one of the items traded.

King Ramkhamhaeng of Sukhothai, Thailand's first kingdom, founded around A.D. 1250, established political relations by sending a mission to China in 1292. Other missions followed and returned to Sukhothai bearing gifts for the King that included a gold-filigreed dress and silk fabrics. They also brought Chinese artisans to improve the production of pottery wares. These artisans may have brought sericulture skills with them.

In 1296, Cho Ta-Kuan, a Chinese envoy, was assigned to a post in Cambodia and spent much of his time in the great city of Angkor. He noted in his journal that, though the Cambodians were not involved in the cultivation of silk, Thai settlers were growing mulberry trees, raising silkworms and producing silk cloth. Cho further explains that Cambodian women were entirely ignorant of sewing, dress-making and mending, and that the dark damask silks worn by Cambodians were produced and mended by Thai weavers. The Thai settlers are believed to have travelled from Sukhothai. This record along with archaeological evidence indicates that sericulture was an established craft of the Sukhothai people. Silk became an item of great value that was used for trade with neighbouring kingdoms such as China and Cambodia.

Sukhothai stone inscriptions record that after the harvest, men made iron implements and women wove cloth. The inscriptions also describe a five-coloured cloth, believed to be silk, used for ceremonial occasions. In the early northern kingdoms collectively known as Lanna Thai which evolved at the same time as the Sukhothai kingdom, silk appeared in the royal courts. The northern city of Chiang Mai and surrounding villages have temple paintings that depict costumes which appear to be made of silk.

The 16th and 17th centuries saw Europeans voyaging throughout Asia and visiting Siam. By this time, silk cloth was firmly established as a

valuable trade item, and it was bartered alongside other Thai goods including ivory, leather, acacia, sapan wood, ceramics and pepper. Siam and the neighbouring courts of Burma, Laos, and Cambodia were renowned for their sumptuous brocades that dazzled the eye with glittering gold and silver yarns. Many silks were woven in India based on Thai designs and imported into Siam. Other silks were Japanese, Persian and Chinese. Records from the court of Ayutthaya, the capital of Siam from 1350 to 1767, explain the techniques of raising silk worms and describe the abundance of the luxurious fabric, used not only for fine clothing but also for wall hangings, dividers and floor spreads. The leader of a French embassy was delighted by silk sarongs of "an extraordinary beauty... permitted to those only to whom the Prince presents them."

It is recorded that in 1608 King Ekatotsarot of Ayutthaya sent a Thai emissary to the Netherlands bearing valuable gifts for the Stadholder, including silk fabrics. Historical records mention that Persian ikat *(mudmee)* worn by a Thai ambassador to the French court in 1685, was so admired that it inspired ikat style weaving and design at the great silk centre of Lyon. Gifts to Louis XIV (1643-1715) from King Narai the Great included costly silks. Unfortunately, when the Burmese destroyed Ayutthaya in 1767, many records and historical details were lost forever.

When Bangkok became the capital city in 1782, many Ayutthayan customs were reinstated. For example, certain Thai silk designs indicated the rank and status of court officials. This distinction based on weave design was practised until the early 20th century. Sir John Bowring, a British consul based in Bangkok during the reign of King Mongkut (Rama IV) noted that costly garments worn by persons of high rank were woven in their own houses, proving that they could produce high quality silks. In 1857, King Mongkut showed Europeans that the quality of Thai silk matched that of fabrics from the British Empire, by sending Queen Victoria magnificent gifts including elaborate brocades, a red silk cloth and a red and gold sarong.

King Chulalongkorn favoured Western-style dress, but often wore a jacket and shirt combined with a silk pha nung chung kraben. *A keen photographer, he is seen holding one of the first cameras in Thailand.*

Cotton was woven in the villages for everyday purposes, but special occasions such as ordinations, weddings and festivals required the use of silk garments. Despite the intricate role Thai silk played in high society, a flood of imported fabrics including fabulous silks from China, Persia and Japan made it difficult for the local silks to compete. Thai sericulture remained a small cottage industry, most active in the northeast region around Khorat until the mid-20th century.

In 1901, King Chulalongkorn made an attempt to upgrade the local silk industry by inviting a team of Japanese experts to aid production. In 1903, the Department of Silk Craftsmen was established under the directorship of Prince Phephatanaphong. These early steps marked the beginning of rapid sericultural development in Thailand. Mulberry trees were planted in the Northeast, local silkworms were cross bred with the Japanese variety, modern spinning and hand reeling machines were introduced and traditional looms were replaced by more advanced ones. Sericulture courses were taught throughout the Kingdom. By 1910, over 35 tons of silk were being exported annually.

Silk production gradually decreased over the following three decades due to a lack of government support and because of strong competition from foreign silks. An inability to improve antiquated techniques resulted in limited production. Only a few companies such as the Shinawatra Silk Company, which had been commercially producing silk since the 1930s, continued weaving and selling silk locally.

AN INDUSTRY REBORN

The Thai silk industry boasts two legendary pioneers who were largely responsible for revitalising Thailand's valuable craft. Chiang Shinawatra, a Thai national, kept silk production going during the industry's hardship years. Later, an American named Jim Thompson brought world-wide recognition to the fabric.

Chiang Shinawatra, the founding father of the Shinawatra Silk Company, began his career in textiles as a cotton weaver in the early 1900s. Shinawatra travelled frequently to Burma, trading fabrics. During his travels, he observed that Burmese silk-weaving operations were superior to those in Thailand. Impressed by what he saw, Shinawatra decided to try his hand at

silk production. Initially, he purchased silk yarn from Burma as his Buddhist ethics made the raising of worms in order to kill them after their endeavours rather distasteful. He experimented with weaving looms and dyeing processes, becoming so successful that in 1939, he registered the company and established his first factory in Chiang Mai. Today, Chiang Shinawatra's descendants manage one of the largest and most sophisticated sericultural operations in Thailand. Operating as a family business, the company employs some of the most advanced breeding practices in the country.

After the Second World War, American architect Jim Thompson came to Thailand with the U.S. military. A fascination with the country and its culture encouraged Thompson to remain in Thailand permanently. His interest in silk was aroused when he collected several pieces and was delighted by the lustrous colours and textural "humps and bumps". Fired with enthusiasm as he recognised the market potential of silk, Thompson started experimenting with a handful of Muslim weavers. His highly successful results led to a revival of the ailing silk industry.

Thompson reactivated sericulture, updated production methods, introduced chemical dyes and created modern designs. His efforts and dynamic business sense were largely responsible for creating a world-wide interest in Thai silk. In 1948, he founded his own silk manufacturing company, catering to both Thai and foreign buyers. Thompson mysteriously disappeared in 1967 while on vacation in the Cameron Highlands of Malaysia. His company, however, remains one of the most respected in the industry, and his example has been widely followed by other entrepreneurs.

The efforts of Chiang Shinawatra, Jim Thompson and later, Her Majesty Queen Sirikit of Thailand and others actively involved in promoting silk, have enabled Thailand to reclaim a prized art that has been practised for centuries. Today, the Kingdom proudly boasts a thriving silk industry.

American entrepreneur Jim Thompson was one of Thailand's 20th century silk pioneers. When silk production was in danger of remaining a small cottage industry after the Second World War, he created such an interest in the material overseas that it led to a revival of silk weaving.

Left: *The yellow cocoons found in the silk weaving villages of Thailand are placed in a pot of boiling water and the silk is reeled by hand. This is a very time consuming process and requires considerable patience and skill.*

Right: *Thai silk is always handwoven, even by the most sophisticated silk producing companies. An experienced weaver may average nine metres per day of a single coloured plain weave.*

From Cocoon To Loom

The creation of a single length of Thai silk is an intricate and time-consuming process. Due to Thailand's favourable climate, sericulture is a year-round activity. After the rice harvest, the villagers of the Central Plains and the Northeast devote their time to silk production. Young girls are taught weaving skills by their mothers as they sit at looms beneath the traditional Thai wooden houses. Their exquisite creations may provide extra income or may be used in their own marriage ceremonies or for funeral rites, or donned during a festival to attract a young man. Weaving is a sign of maturity and eligibility for marriage.

Silk cultivation is predominantly a cottage industry, in which villagers have varying degrees of involvement. Some grow mulberry trees and only raise

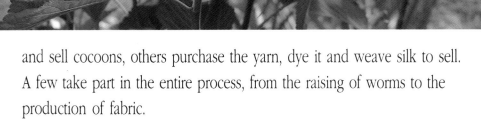

Above: *The quality of the mulberry leaves either of the cultivated or wild variety affects the quality of the silk yarn because the fastidious silkworm thrives only on the very best, which must be freshly picked and chopped several times each day.*

Right: *Non-stop eating machines, young caterpillars are very sensitive to noise, odours and disease. A box containing 20,000 caterpillars would consume around thirty sacks of mulberry leaves each day.*

and sell cocoons, others purchase the yarn, dye it and weave silk to sell. A few take part in the entire process, from the raising of worms to the production of fabric.

The origins of the shimmering, exquisite fabric lies in a caterpillar that belongs to the insect order *Lepidoptera*, which includes all moths and butterflies. There are two distinct types pertinent to silk production: the *Bombycidae*, or cultured silkworms, and *Saturniidae*, the wild silkworms.

The Bombyx *mori* of the *Bombycidae* family is the most common source of cultured silk. The worm is raised domestically, but only where mulberry leaves are available to satisfy its voracious appetite. The wild silkworm species, of which there are more than 500 types, will devour other varieties of leaves. More robust than their domesticated cousins, *Saturniidae* produce a tougher, coarser silk which is shorter in length, with colours that vary from off-white to beige or yellow. Wild silk is usually called tusser-silk. In Thailand, the two predominant Bombyx *mori* varieties used are the bivoltine, which produces the fine white yarn suitable for power-loom weaving, and the traditional hardy polyvoltine, which produces the irregular yellow thread suited to hand-loom weaving that is used in the creation of *mudmee*. Bivoltine silkworms produce two harvests a year, and polyvoltine an unlimited number.

The silk caterpillar goes through four stages in its life cycle: egg, larva to caterpillar, pupa or chrysalis and moth.

From a box containing 20,000 eggs, it is expected that 15,000 cocoons will be reared successfully. To minimise losses, the cocoons are divided into separate trays and closely observed. In an effort to increase silk production and improve the quality of Thai silk, Japanese sericulture is being introduced under technologically advanced and highly controlled conditions.

Rearing of the silkworms determines the quantity and quality of the silk, since they are fragile creatures that need cosseting. Silkworm keepers must ensure that their valuable charges are not only correctly fed, but also kept meticulously clean. Silkworms are highly sensitive to noise, odours and other factors in their environmental surroundings.

Mature silkworm moths mate for several hours, and the female deposits 300 to 500 grey pinhead-size eggs. Two or three days later she dies. During germination, the eggs require cool temperatures for development and warmer temperatures to hatch. Developing and hatching can take between 10 days and 10 months depending on climatic conditions. In Thailand, eggs kept at controlled temperatures of 24 to 28 degrees Centigrade take about 21 days to hatch, whereas eggs in the wild require only 10 days because of the warmer conditions.

The young larvae begin feeding from the moment they hatch. This is the most crucial part of the cycle, since the worms are vulnerable to predators and to silkworm disease. Initially, they require three servings a day of clean, tender mulberry leaves, freshly picked and chopped. Preparing the leaves is labour intensive, round-the-clock work. Gorging for 28 days, the cultivated variety increase their body weight an astonishing 10,000 times. They grow to about three inches, and shed their skin four times. By the end of the feeding cycle, the caterpillar will consume 500 times as much as when it was newly hatched. When fully satiated, the caterpillars raise their heads to indicate the commencement of spinning.

Before serious spinning begins, the worm secretes a pale yellow gum from two glands on either side of its head. From this, it creates a web on which it anchors its cocoon. Once anchored, the worm ejects liquid silk from its two glands. Upon exposure to the air, the secretion becomes a fibre. In spinning the cocoon around itself, a process that takes two to three days, the worm works in a figure-of-eight motion, distributing the gummed thread evenly and winding from the outside to the centre. The cocoon effectively becomes the worm's shroud.

The Thai pupa remains inside its cocoon for about 23 days. It pupates and spins a filament between 550 and 730 metres long. Other

A fully grown caterpillar sheds its skin four times and increases its body weight an astonishing 10,000 times, growing to around three inches in length. It is an oyster-white colour and has a retracting head.

silkworm varieties can produce up to 1,650 metres of filament. To produce a mere 100 grams of silk thread, over 1,000 Thai cocoons are required. Over 500 cocoons are needed to weave a single necktie, around 4,000 a blouse and 8,000 an evening dress.

When spinning is complete, the cocoons are boiled or steamed in order to kill the pupa, to prevent it from metamorphosing into a chrysalis which would break the thread upon emerging from the cocoon. In Thailand, reeling or unwinding the silk filament from the cocoon is traditionally done by hand. After sorting, the cocoons are placed in a pot of boiling water which softens the sericin and kills the pupa. A long-handled wooden paddle with a notch in the centre is used to submerge and stir them. When the cocoons are pressed under the water, the silk threads float to the surface and are swiftly reeled onto a wheel or frame 10-20 filaments at a time. The notch in the centre of the paddle prevents the cocoons from becoming entangled. As the reeling of the filament from each cocoon nears completion, the filament from a new cocoon is attached to it. Acting as an adhesive, the sericin assists in binding it.

Three types of silk are obtained from the reeling process:

Mai ton or *hua mai* silk is obtained from the first reeling or the outermost part of the cocoon. The threads are large, coarse and of a dull yellow shade. After refinement, bleaching and spinning, they become thick, coarse silk threads.

Mai klang or *mai song* threads are obtained by a continuous reeling process. When the filament from one cocoon is finished, it is replaced by another cocoon. After constant reeling, the reeler can adjust the size of the silk filament to medium or smooth with a few knots.

Mai noi or *yod mai* silk is obtained from the innermost part of the cocoon. It is yellow, fine and delicate. Due to the complexity of the reeling process, this is the most expensive type of silk.

After removal from the reel, the skeins of raw silk are soaked in hot water to remove the remainder of the sericin. At this stage Thai silk yarn is yellow. Before dyeing it must be bleached in hydrogen peroxide or lime water and then dried thoroughly in the sun. Silk skeins are transformed into

Right: When it has finished feeding, the caterpillar anchors a strand to the surface and spins a web as a framework for the cocoon — an endeavour that takes two to three days, before the real work begins.

Right: The yellow cocoons are the polyvoltine variety of silkworm which produce the yellow, coarser thread that is used in weaving mudmee *silk. An average cocoon contains around 500 - 700 yards of filament.*

Right: The white cocoons are the bivoltine variety, which have often been imported from Japan. They produce a finer, longer thread which averages between 1200 - 1800 yards in length.

33

silk yarn by throwing, a process where several filaments are twisted together to form a ply. It is the thickness of the finished silk or the ply that dictates the purpose for which it may be used.

Nowadays, most dyes are of the imported colourfast chemical variety. Many villages, however, continue to produce organic dyes and mordant from plants, roots, insect resin and soil. The leaves of a *krarm* tree produce a blue dye, the indigo plant or the lac insect and its nest or raw sealing wax found on the barks of trees are the basis of a red dye. Pink dye may be obtained from the betel nut. An indigenous wood called *kha*, tumeric, or the core of a jackfruit tree, produces yellow, and the berries of the ebony tree make black or brown. Green dye is obtained from the roots of the *talaeng* plant, walnut shells, or the leaves of the pineapple tree. Raw dye materials are crushed or sliced and then boiled to extract the colour. The sediment is removed, and the remaining liquid is suitable as a dye.

Natural dyes produce more subtle colours than chemical ones. Ingredients vary according to regions and are often closely guarded secrets passed from one generation to the next. The skeins of silk are immersed in the dye and left to soak. The dyed yarn is then fixed with a mordant made from plant extracts or of chemical origin. The result is a beautiful and originally coloured yarn.

Even in the most modern of factories, Thai silk is handwoven to avoid damaging the delicate thread. It is this feature that assists in creating its special lustre and distinction. However, with the constant improvement of silk thread quality, power-loom weaving should become more popular in the future. Increased demand for Thai silk has prompted a metamorphosis from a cottage industry to factory production, and has resulted in a shortage of silk warps and wefts. Imported silk yarn from countries such as China or Japan are employed in the warp — the stronger, more even thread stretched lengthwise on the loom. Local yarn is used on the wefts — the lateral threads woven across the loom. The use of imported silk yarn helps to maintain quality and allows standardisation of silks for export.

Various types of traditional looms are used in Thailand. They are usually made from hardwood or bamboo. The most common is the "flying shuttle", popular because it is both fast and simple.

Right: *In the villages of the Northeast where silk production is still a cottage industry, the cocoons are sorted and placed on traditional trays made from bamboo.*

Pages 36-37: *When the silk is reeled from the cocoon, several filaments at a time may be twisted together to form a ply — which determines the thickness of the silk.*

When the caterpillar has finished spinning what in effect becomes its own shroud, the cocoons are boiled in hot water or steamed at a high temperature to kill the pupa to prevent it metamorphosing into a butterfly and damaging the thread as it emerges from the cocoon.

Once the warp is set, the yarns are threaded through a heddle eye, a looped cord to control the movements of the yarn during weaving. Each heddle is attached to a rectangular frame called a harness. Alternate strands are raised and lowered by the heddle frame using a foot pedal, so that the weft, or filling yarns, alternate in passing under one set of warp yarns and over another. The weaver pulls on a shuttle cord to send the shuttle flying to and fro across the warp to fill in the yarn. At the same time a reed, a device to separate the warp threads and guide the shuttle, moves back and forth pushing the newly woven weft threads tightly into the weave.

A plain weave requires two heddles. Each weft yarn goes alternately under and over the warp yarns across the width of the fabric. On each pass, the weft yarn alternates the interlaced pattern. Intricate designs or brocades require a large number of heddles. To form the elaborate patterns that are so popular for ceremonial occasions, gold and silver brocade weft threads are slipped in and out of the warp. When a design is being planned, a pattern of the weave has to be sketched out in order to set up the loom correctly and to indicate the particular heddle through which each warp yarn is to be drawn.

In the cottage weaving industries, and especially in the production of *mudmee* tie-dyed silk, hundreds of individual patterns are achieved by using different weaving and dyeing techniques. Designs are based on elements of nature such as birds, flowers, fruit, snakes, trees and animals or religious

Right: *After reeling and before the dyeing process, the raw silk threads have to be washed to remove the sericin and dried out.*

Right: *As much Thai silk is of the yellow variety, it has to be bleached to white before it can be dyed in other colours.*

Right: *After dyeing in chosen colours either using natural dyes or the chemical variety, the silk threads are hung up to dry before being twisted into skeins.*

Pages 40-41: *The results of bleaching and dyeing are rich, gleaming colours that are ready for winding and weaving.*

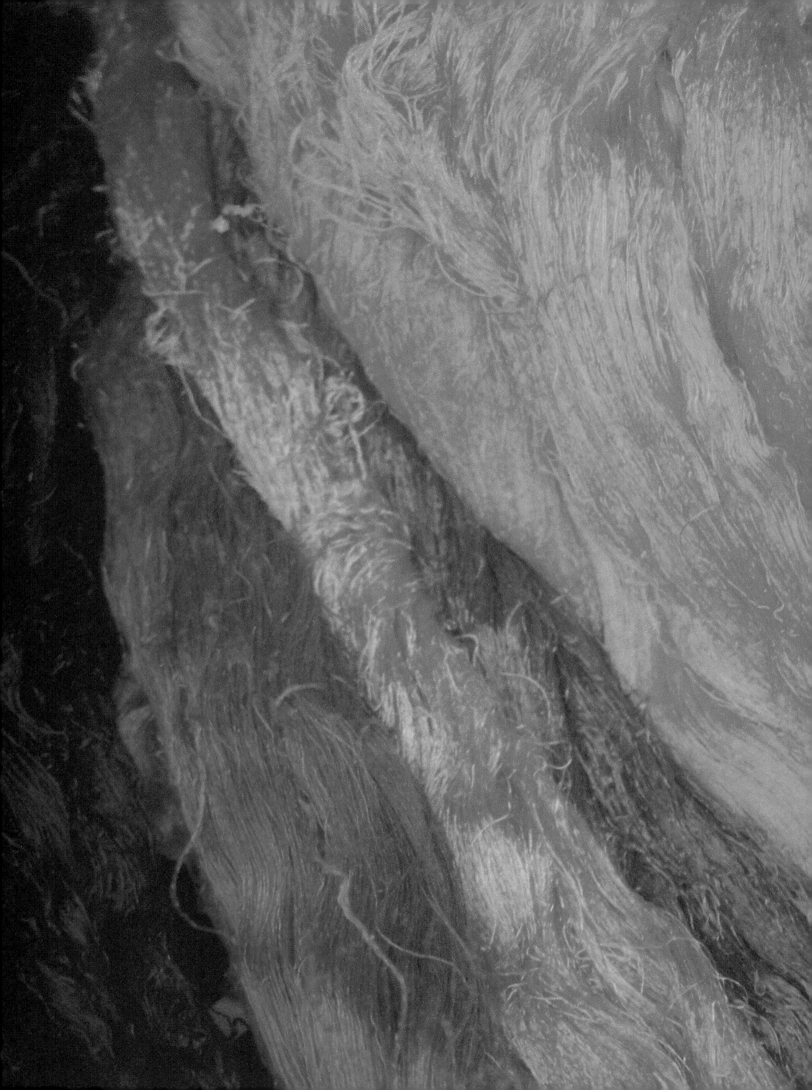

symbols. Many are universal, common throughout the Kingdom. Others are unique to particular villages. Weaving requires a high degree of skill. An experienced weaver averages about two metres of woven patterned fabric or nine metres of single coloured silk in a day.

Although computerised pattern printing technology is popular abroad, the larger operations in Thailand use mechanised silk screen printing methods. The results are equally sophisticated. In the flat screen printing technique, a design is copied onto a number of very fine flat screens, one for each colour to be printed. The design for each screen may be drawn by hand or stencilled, or a photographic negative may be used. An impermeable substance is applied to those parts of the screen that are not part of the design, to prevent colour being absorbed. Colour samples are mixed and tested on small pieces of fabric to ensure that they match the design.

Once the paints, print heads and fabric have been prepared, the silk is attached to a backing spread and laid flat on a long printing table. A screen representing one colour of the design is set above the fabric. A printing carrier holding one pattern screen at a time is pushed to and fro the length of the fabric to apply one colour at a time. Once each colour has dried, the screen is changed and the process repeated until the pattern is complete. The printed fabric is hung on a rack above the printing table to dry. Thirty-six metres of continuous silk, with nine individual colours, requires an application time of five hours.

In automatic screen printing the process is faster, as electronic controls set each screen on the fabric. The colours are released as the fabric automatically moves along a conveyor belt to the next screen, adding different parts of the design as it progresses. When the pattern colouring is complete, the finished product passes into a drying machine. The advantage of this method is that up to 20 screens may be employed for as many colours.

A few small printing operations still use the hand-block printing method. The colour and pattern are applied by stamping the pattern onto the material by hand, an ancient art that produces splendid designs. Carved blocks may be made from stone, wood or other suitable substances.

Following the printing process, the silk is steamed to fasten the colour. The fabric is then washed in boiling water in order to rinse out any

The dyed silk strands are wound onto spools and at this stage the thickness of the plies may be adjusted.

unfixed dyes or adhesive. After a final drying, chemicals are applied to the material to prevent shrinkage, minimise creasing and ensure easy care.

The results of meticulous spinning, dyeing, weaving and finishing efforts are manifest in exquisite pieces of Thai silk.

Pages 44-45: *The final part of the process is weaving. The warp threads are set and the creation of a length of silk materialises.*

Above: *In creating individual patterns for special silk products such as high fashion, the hand painting technique is used to apply a pattern.*

Left: *Using a traditional loom made of bamboo and hardwood, producing two metres of patterned weave may take an experienced weaver a minimum of one whole day.*

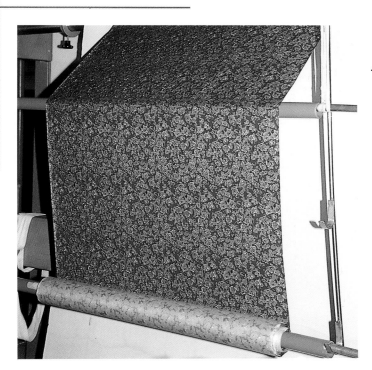

A	B
C	D

Left: *In large commercial silk operations, where long lengths of the same design are required, the pattern is produced by silk screening and applied by table printing.*

A: *After the pattern has been applied, the silk is steamed to fasten the colour.*

•

B: *The fabric, which may average around forty yards in length, receives a final application of chemicals to fix it.*

•

C: *The silk is dried as it passes along a conveyor.*

•

D: *The finished fabric is stretched, pressed and wound into sheets of patterned silk.*

Left: *King Chulalongkorn (Rama V) together with one of his consorts, Queen Sukhumal. The King is dressed in silk stockings and a silk* pha nung, *and his consort is wearing a Thai-style skirt that is pleated in the front and a* sabai, *a loose cloth wrapped like a sash across the upper body and worn over a European blouse.*

Right: *A Thai princess in the reign of King Vajiravudh (Rama VI, 1910 - 1925) looks coyly elegant in a Victorian lace blouse and a Thai-style* pha nung *made from Indian silk in the floral design used by the Royal Court.*

On Royal Command

Silk and fine textiles have always held an important place in the Thai royal courts. Magnificent brocaded silk costumes, heavily embellished with gold, were a sign of wealth, power and noble birth. They enriched ceremonial occasions and served as a medium of exchange or gifts. A knowledge of costumes and their importance can be gained from mural paintings and figurines or from descriptions by early foreign travellers.

Records indicate that during the Sukhothai period (around A.D. 1250 - 1438) and Ayutthayan period (A.D. 1350 - 1767), exotic silks, brocades and satins were imported from China, Persia and India to make elaborate court costumes for the nobility. Aristocratic ladies wore a court *pha sin,* a simple garment one metre long, gathered and folded at the waistline and secured with a belt. This was combined with a *sabai,* a long piece of plain pleated or patterned silk about 30 centimetres wide that was worn above the

Her Majesty Queen Sirikit of Thailand wears the Thai Chakri costume. The skirt has a pleated central panel and is held together with a golden belt. The sabai is heavily embroidered and worn over the right shoulder.

waist, draped across the chest, falling casually over the left shoulder. The fabric could be worn with either one or both ends tucked into the waistband, or secured by small gold chains. The *sabai* was often scented with perfumes made from a potpourri of flowers, fragrant limes and incense.

Foreign visitors to the court of Ayutthaya always mentioned the glittering beauty of the clothing. Simon de la Loubere, a French emissary, visited the city in 1688 and provides the following description: "They go with their Feet naked and their head bare; and for Decency only they begirt their Reigns and Thighs down to their Knees with a piece of painted cloth about two Ells and an half long, which the Portuguese do call *Pagne*, from the Latin word *Pannus*... sometimes instead of a painted cloth, the *Pagne* is a silken Stuff, either plain or embroider'd with a border of Gold and Silver."

The *pha nung*, a rectangular piece of material about one metre wide and three to four metres long, was popular amongst both men and women. It could be worn in two different styles: as a *pha nung chung kraben,* Indian *dhoti* style, or as a *chip-na-nang*. In the *pha nung chung kraben* style, the garment was tied in the front, knotted at the waist, then twisted from the top edge to the bottom edge to form a tail which was folded between the legs and tucked into the waist at the back. It was held in place with the aid of a belt or loincloth. In the *chip-na-nang* style the garment was wrapped around the lower body, gathered in front with both ends skilfully folded into pleats and fastened with a belt.

In addition to the complications of dressing properly, Ayutthayan law governed fashion propriety. Laws dictated that people should dress according to their social class and decreed that certain dress styles would be restricted to royalty or members of the court. Only the Queen was permitted to wear the *hom-pak,* an ornate heavily jewelled type of *sabai*. Commoners went bare chested, wearing a simple, wrap-around skirt-like garment.

In the 18th century Ayutthaya was attacked by the Burmese, and the capital finally fell in 1767. During the period of warfare, women were forced to crop their hair short and disguise themselves as men so as to avoid being captured and carried away by the Burmese. Elaborate styles were abandoned for the sake of practicality. The long and confining *pha sin* skirt, which would have handicapped a woman needing to defend her family, was replaced by the *pha nung* worn with the *ta-beng-man,* a crossed halter top.

Her Majesty Queen Sirikit is Thailand's best ambassador for Thai silk. To keep traditional Thai styles of dress alive, the Queen designated five styles to be worn for different occasions by Thai ladies.

Two older aristocratic ladies during the reign of King Mongkut (Rama IV, 1851 - 1868) wear pha sin *skirts and silk* sabais *thrown across the shoulder.*

Right: *Queen Saovabha, the consort of King Chulalongkorn (Rama V) poses with three of the King's many children, who also combined Western and Thai styles of dressing in silk.*

Once the new capital was established in Bangkok and conditions stabilised, women turned their attention to fashion. The Rattanakosin period (late 18th century to early 20th) saw those styles prevalent in Ayutthaya revitalised. Though men did not wear coats or tops unless duty required, women recommenced wearing the *sabai*. *Pha sins* became richer and more colourful and were embellished with ornate jewellery.

During the 19th century, Thailand expanded contact with the West, and was influenced by European and American culture in all realms of life, including fashion. During the reign of King Rama III (1824 - 1851), a visiting American diplomat described the dress of a royal prince: "He was dressed in a jacket of pink damasked crape, closely fitting the body, a sarong of dark silk, knotted in front, the two ends hanging down nearly to the ground, and over it was tied a light sash, upon which two jewelled rings of large size were strung. This costume left the head, arms and legs bare..."

King Mongkut (Rama IV, 1851 - 1868) favoured and encouraged Western styles of dressing. The monarch and aristocratic men began to wear long-sleeved tops and pants, while high society ladies wore Western-style skirts together with blouses. When attending royal functions, in the presence of the King, officials were required to wear jackets, and uniforms were issued according to rank.

King Chulalongkorn (Rama V, 1868 - 1910), influenced by a visit to Europe at the end of the 19th century, followed in his father's footsteps in the westernisation of clothing. However, during special occasions, noblemen and officials reverted to traditional Thai costume. Men would often combine a *pha nung* with a Western-style jacket, shoes and stockings. Photographs of the King show him wearing either trousers or a *pha nung*, depending on the occasion. The King enjoyed travelling around the country accompanied by his consorts, whom he prefered to be attired as femininely as possible. He objected to the masculine garb popular amongst contemporary travelling women. Thus, traditional Thai dress was redesigned to please and accommodate the "modern" Thai woman. The *sabai* was now worn over a blouse, and the *pha nung* was frequently worn with a silk jacket trimmed with lace, or with a camisole top with bows along the straps, or a high-necked Victorian style blouse and a piece of colourful silk draped like a sash over one shoulder.

J.G.D. Campbell, a foreign traveller visiting Bangkok in the late 19th century, eloquently described the clothing of the era: "The national garment of the Siamese of both sexes is called the 'panung' .... the costliness of the material used, cotton or silk, varies with the means of the wearer. The 'panungs' of the rich, being often of beautiful silk are very handsome, their very simplicity adding to the effect. They can be seen in all hues — orange, green, blue, red and purple in every shade — each day of the week, it is said, having its appropriate colour. "He further wrote that the wealthy wore long silk stockings and often patent leather shoes with buckles, and that wealthy men wore a white jacket and ladies of the upper class silk or satin embroidered jackets trimmed with costly lace and jewelled buttons.

Styles also varied geographically. For example, in the northern court of Chiang Mai which became a province of Thailand under Rama V, women never wore the *pha nung chung kraben*. Instead, those of high rank wore a plain or brocade *pha sin*, a skirt folded in front, with an elaborately embroidered silk or cotton hem interwoven with gold or silver thread.

When King Vajiravudh (Rama VI, 1910 - 1925) announced his betrothal, he also made it known that he wished women to discard the *pha nung* in favour of the *pha sin,* which gradually gave way to European-style skirts. During his reign, city women preferred the new fashions, while many country dwellers continued to wear the *pha nung*. However, some older ladies

Above: *Queen Saovabha, the consort of King Chulalongkorn, combines an elaborate French lace blouse with a silk* pha nung, *lace stockings and a* sabai *with a decorative bow on the shoulder.*

felt undressed without a *pha nung chung kraben* and chose to maintain one beneath their skirts. The *sabai* virtually disappeared, but the *pha sin* regained popularity. Mixing Eastern tradition with Western styles, the *sin-yok,* a cloth woven in gold or silver was frequently worn long over loose, Western-style blouses.

With the introduction of constitutional monarchy in 1932, the *pha nung* was deemed improper and was replaced by Western-style uniforms. This change reflects the extent to which traditional Thai costume has been mixed with or replaced by contemporary Western fashion.

In an effort to preserve the heritage of Thai dress, Her Majesty Queen Sirikit decreed five principal styles of national costume for wear by modern Thai women. Derived from the traditional dress of earlier periods, they are categorised according to levels of formality.

The **Thai Ruanton,** meaning Thai-style house, is a casual dress suitable for informal day wear. The skirt is long with vertical or horizontal stripes on the hem, while the jacket has three-quarter sleeves with a round, collarless neck. Formal daytime occasions call for the **Thai Chitrlada,** named after the palace which is the residence of the Royal Family. This is a long-sleeved jacket with a stand-up collar and a *pha sin* skirt with a centre pleat either plain or brocaded.

During an informal evening function, one should wear the **Thai Amarin,** an outfit of skirt and jacket named after the most beautiful throne room within the Grand Palace complex. This skirt material may be of silk brocade. The jacket may have either three-quarter length sleeves and a collarless neck or long sleeves with a stand-up Thai collar. Formal evening occasions require the **Thai Borompimarn,** meaning great heavens. This is a one-piece dress created from gold or silver brocade. The skirt is pleated down the centre panel, and the bodice has long sleeves, a high, round neck with a stand-up collar.

The full dress costume worn for formal palace ceremonies is called the **Thai Chakri**, meaning palace or dynasty. This is a two-piece outfit, with a skirt of heavy gold or silver brocade and a pleated centre panel held together by a silver or gold belt. The top or *sabai* may be attached to the skirt or worn separately covering one shoulder.

Left: *The Thai Borompimarn has been designated suitable for formal occasions.*

Above: *A modern variation of costumes worn by Thai ladies in the 19th century where Western blouses were combined with Thai* pha nung *made from silk.*

Pages 60-61: *The* sabai *may be worn over long or short sleeve blouses and may be plain, patterned or brocaded.*

Right: *Early Bangkok costumes were varied according to the colour of the* sabai *and the skirt, which was gracefully gathered at the waist with a gold belt.*

Above: *During the 18th century at the fall of Ayutthaya, Thai women opted for practicality in their dress and wore a simple crossed halter top called a* ta-beng-man *and* pha nung.

Pages 64-65: *For daywear, the Thai Chitrlada and Thai Ruanton are elegant two-piece outfits which can be made in many different colours.*

Left: Mudmee *silk is characterised by its unique iridescent colours and fuzzy edged patterns.*

Right: *Queen Sirikit of Thailand frequently wears beautiful* mudmee *clothing made from silk produced by weavers from the SUPPORT Foundation.*

Magnificent Mudmee

Thailand's northeastern provinces are home to a fascinating method of textile production, a tradition which has been practiced for centuries. It is known as *mudmee* silk weaving or when translated, tie and dye silk weaving (ikat).

Mudmee silk is characterised by unique iridescent colours, original designs and fuzzy-edged patterns. Traditional *mudmee* colours are based on hues of black, red, brown and yellow, although each region possesses its own variations. Because *mudmee* weaving involves far more human craftsmanship than machine work, the creation of *mudmee* silk is an art rather than an industry; no two lengths of fabric are ever identical. A difference exists in the thickness of the thread, the subtle shade of the dye or a slight variation in two patterns that, without looking closely, appear superficially the same.

Lengths of silk yarns are stretched across a lakmee *frame and the pattern is tied on by hand, a time-consuming process that requires great patience.*

Left & Above: *Nylon chords are tied onto the strands of silk yarn to form the pattern and to control the placing of the dye.*

Many silk-producing countries weave their own variation of ikat. However, the ikat of each country is distinguished from that of others by the ingenuity of the local weavers in the use of dyeing techniques and pattern designs.

A *mudmee* design is created by a complex dyeing system. The designer must envisage the colours of the fabric and the pattern required and then work with this image in mind while tying and dyeing the yarn.

The warp, the weft, or in some cases both the warp and weft yarns are dyed only in selected areas when intended for use in *mudmee*. The yarn is strung across a frame equal in length to the size of the fabric. The number of yarns required for the pattern are counted before being moved to another frame where the tension of the stretched yarn can be adjusted.

Each stretch of yarn is individually tied by chords made from the dried bark of a banana tree or by plastic threads. A tied area will resist the absorption of pigment when the rest of the strand is dyed. However, a subtle blurred line occurs at the edge of a tied area where the dye has bled.

The tied yarn forms an outline of the intended pattern. After the first tie and dye process, the yarn is untied and dried. The production of a multi-coloured fabric requires that the tie and dye process be repeated several

Above & Right: *After the chords have been tied, the silk is dipped into the dye mixture and the process repeated for each separate colour.*

Pages 72-73: *Before applying the next colour, the silk must be hung out to dry.*

times. Once all tying, dyeing and drying is complete, the yarn is wound onto spools that are marked and stored in the correct pattern weaving order. Misplaced spools can ruin the entire design process.

Mudmee patterns are commonly derived from themes based on nature. Designs encompass all types of flora and fauna, and have been passed down through generations. Unfortunately, many of the old designs have been forgotten or lost because they were never recorded. Most weavers and dyers carry and plan patterns in their heads, although they commonly vary shades of the same colour or borders. Simple designs can be achieved with the use of only three or four colours, whereas the more elaborate might incorporate eight or ten.

One of the simplest *mudmee* designs is known as "falling rain", where odd threads of the weft and sometimes the warp have been randomly wrapped to give small flecks of light colour after dyeing. Many *mudmee* patterns are based on the use of lines and depict bamboo leaves, snakes or diamond shapes. Geometric patterns are common. A design often seen is a diamond-shaped pattern scattered diagonally across the cloth which may be arranged in widely spaced formations or close groups. Patterns commonly found in the northeastern areas of Surin and Khon Kaen favour bird and animal motifs, such as elephants, spiders, turtles and butterflies, that are incorporated into the borders of the silk lengths; Surin is especially renowned for its fine elephant and peacock designs. Some of the villages around Udon Thani and the Mekong River base their designs on those of Laos. Part of the originality of *mudmee* is that the same pattern can appear very different when unusual colour combinations are used. When viewed in strong light, the nuances of the colours continually dance and change.

The craft of weaving *mudmee*, known as the "Queen's Silk" because of its association with Her Majesty Queen Sirikit, had almost disappeared two decades ago under competition from imported or factory produced fabrics. Village women wove a few pieces that were kept for their own use on special occasions, since it was not a commodity that they considered commercially viable or of great interest to anyone.

The unique beauty of the *mudmee* patterns and colours were rediscovered by Queen Sirikit on a visit to the Northeast in 1970. She travelled around various villages talking to weavers and collecting old pieces of silk — some of the best, Her Majesty discovered being used as floor cloths. After selecting patterns that she felt were especially attractive and would be marketable, she asked the weavers to copy them and ensured that they were paid a fair price. Her Majesty bought much of the finished silk herself, and the royal wardrobe soon included stunning outfits in *mudmee*. Fashionable ladies immediately followed suit, and *mudmee* gained a following in high society.

In 1976, the Foundation for the Promotion of Supplementary Occupations and Related Techniques (SUPPORT) was established under the patronage of Her Majesty Queen Sirikit to help carry out her many handicraft projects that provide a source of income for the rural poor. Queen Sirikit takes an active personal role in the textile production, often inspecting many of the *mudmee* pieces and making recommendations on pattern designs and quality control. A number of young girls are fortunate enough to be trained at the SUPPORT centre within the grounds of Chitrlada Palace, the Bangkok residence of the Royal Family. Upon completion of training, they return to their villages equipped with skills to support their families.

Today, over 7,000 weavers work under the SUPPORT project and are well paid for their labour. Due to the efforts of Her Majesty Queen Sirikit, overseas fashion shows featuring designs in *mudmee* are frequently held, and the fabric is sold in many countries.

The Chitrlada shops were established as retail outlets to sell *mudmee* silk cloth made specially by SUPPORT weavers. These shops are located throughout Thailand, including those in the Grand Palace, Vimanmek Teak Palace, Bangkok Airport and Bang Sai Art and Crafts Centre near Ayutthaya. The silk fabrics sold at these locations are of the highest quality. Only the best pieces are selected, so buyers are assured of a handwoven piece of art.

Note: Contemporary Thai silks from various provinces and woven under the SUPPORT project may be viewed at the SUPPORT Museum in the Apisek Dusit Hall, a lovely old building that was once utilised by Parliament and lies close to Vimanmek Teak Palace and the Ananta Samakhom Throne Hall.

Queen Sirikit frequently makes trips upcountry to personally check the work of SUPPORT weavers and offer suggestions on design.

Pages 74-75: Mudmee *silk is always handwoven, on wooden looms set up beneath the house such as in this northeastern Thai village.*

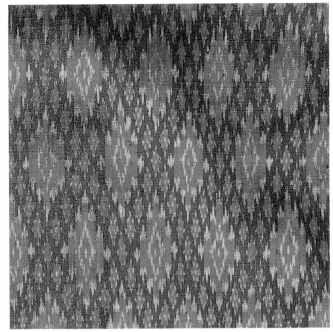
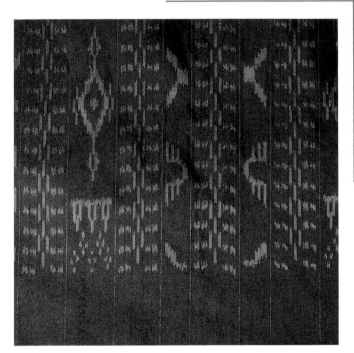

Mudmee *patterns are extremely imaginative. They are inspired by themes from nature, Buddhism and animals.*

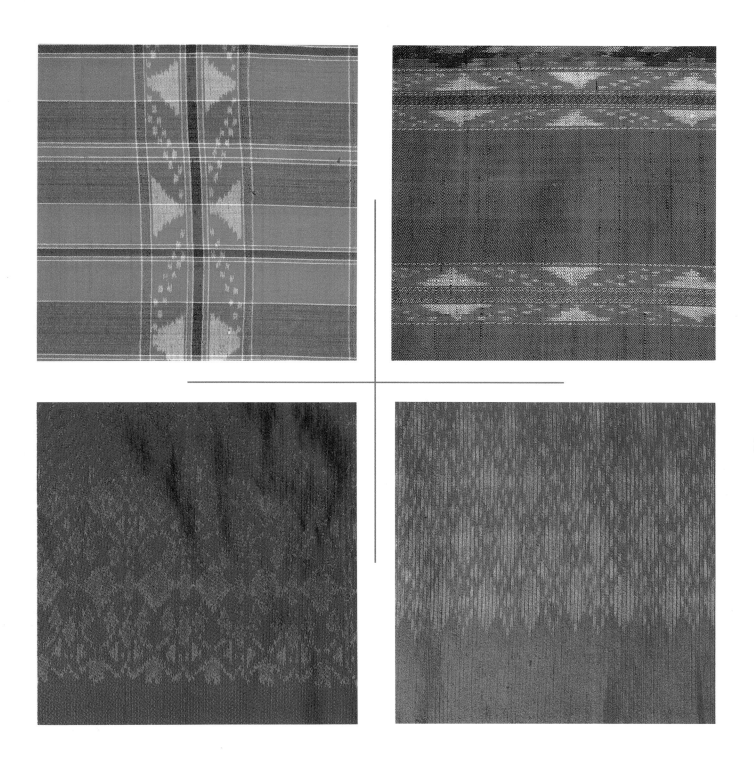

No two lengths of mudmee *silk are ever identical due to the process of tying and dying the pattern and weaving by hand.*

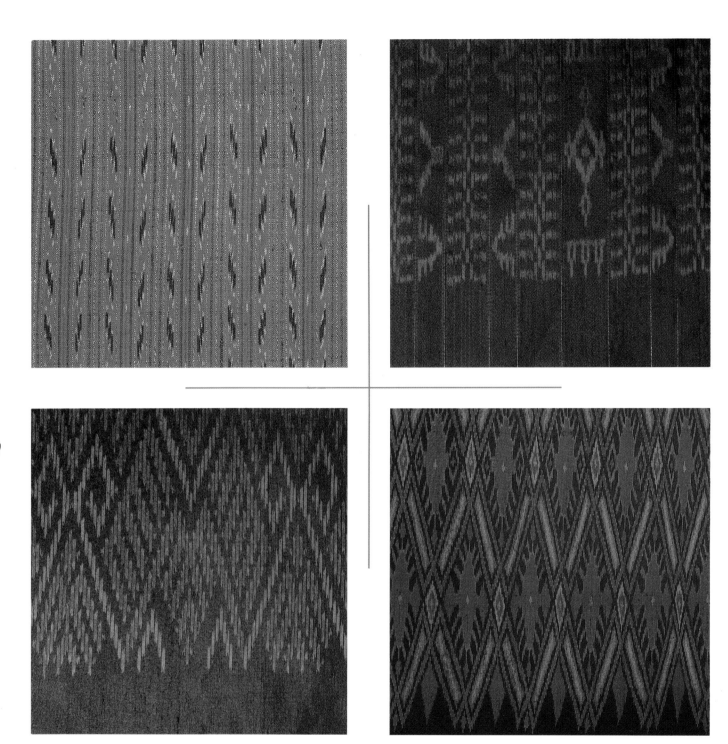

Although chemical dyes are widely used, a few weavers still prefer to make natural dyes from roots, berries, trees and forest products. The results are incomparable colours and lengths of silk that are works of art.

Known as the "Queen's Silk" lengths of mudmee silk are a combination of skilled craftsmanship and hard work.

Left: *Thai silk inspires fashion designers to create beautiful, bold and original clothing, such as this dress and jacket by top designer, Kai.*

Right: *Aristocratic Thai ladies were very fashionable during the 1920s such as this young lady who shortened her Thai silk wrap-around skirt and combined it with a Western-style blouse and shawl.*

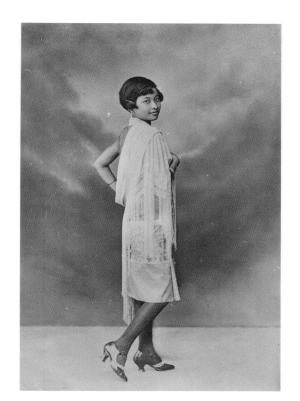

The Fashion World Discovers Thai Silk

Silk. To fashion designers, the very name evokes extravagant lustre and splendour. An exciting, incomparable work of textile art whose soft rustle conjures up images of an unmatched elegance that over the centuries has graced kings and queens, emperors and empresses. After the Romans discovered silk, Julius Caesar sported royal purple silk outfits every time he appeared in public, while in Egypt his paramour Cleopatra was equally enamoured of the fabric.

In the early 1960s, Her Majesty Queen Sirikit travelled to the United States, taking with her a splendid wardrobe of Thai silk designed by Pierre

Balmain. Her voyage marked a trend-setting turning point for Thai silk in the West. Although Thai ladies had been gracefully attiring themselves in traditional silk costumes for centuries, Westerners tended to regard Thai silk as an exclusive fabric used only during formal occasions. Through Queen Sirikit's example, and with the creative brilliance of both foreign and Thai designers, Thai silk is now used as a medium for everything from evening gowns to handkerchiefs.

The American entrepreneur Jim Thompson is generally deemed responsible for stimulating world-wide interest in Thai silk. In 1946, he took some choice samples to New York, where they were publicised in *Vogue* magazine. Top designers immediately bought the silk and transformed it into dazzling dresses. Thai silk, however, was at that time only available in small quantities and was in competition with the more celebrated Chinese, Japanese and Indian silks.

In 1948, Thompson founded the Thai Silk Company, thereby beginning the task of revitalising Thailand's silk industry. He enthusiastically promoted the product to tourists by standing in the lobby of the Oriental Hotel with pieces of silk draped over his arm, enticing visitors with the rich shimmering fabrics. They found the exotic, original colours hard to resist. Thompson's initial success led to the opening of his first small shop in Bangkok, on Suriwongse Road. As demand for Thai silk increased, the Thai Silk Company appointed agents abroad to attract the attention of designers in America and Europe who were always looking for something unusual and original. Due to Thompson's endeavours, Thai silk received an extra boost when used for costumes in the film version of the Rodgers and Hammerstein musical "The King and I" and later, in "Ben Hur". So much publicity was generated that orders rapidly began to flow Thompson's way.

A growing demand from tourists for Western-style garments made out of Thai silk inspired Vera Cykman to establish Star of Siam in 1956. Hers was one of the first companies in Thailand to manufacture ready-to-wear Thai silk apparel. Cykman worked in collaboration with Jim Thompson. She took silk produced by Thompson's weavers and transformed it into stunning suits, dresses and blouses. She opened her first boutique under the staircase of the Oriental Hotel where it remained for 33 years.

Opposite (top): *Bergdorf Goodman, New York, was one of the first overseas stores to sell ready-made Thai silk designs in the early 1960s. The first items to go on display were made by Star of Siam.*

Opposite (below): *Exquisite silk designs by the Star of Siam caused tremendous interest in the fashion world when they went on sale in the United States.*

Pages 86-87: *The early 1960s saw the arrival of Thai silk fashion on the international scene (Star of Siam).*

Star of Siam's designs were an enormous success. They soon appeared in major department stores in America and Europe and were publicised in magazines such as *Harper's Bazaar, Vogue* and *Women's Wear Daily*. Today, the company remains a fashion leader in chic, ready-to-wear silk garments, and has several outlets around Bangkok.

In 1979, the famous French designer Pierre Balmain unveiled his spring and summer collection at a charity fashion gala in Bangkok. Seven elegant and distinctive evening creations of *mudmee* silk were the highlight of his "haute couture" collection. From that time on, Western and Thai designers began to use Thai silk to create garments for casual wear. Once again Queen Sirikit set fashion trends; in 1985, on a visit to America, she wore a wardrobe designed by Eric Mortenson of Pierre Balmain. Both day and evening wear were created from *mudmee* silk which dazzled and impressed her American hosts.

The Minaudiere Moment-a clutch of silk from Thailand... At-Home Collections, Second Floor.

Above and Right: Vogue, Women's Wear Daily *and* Harper's Bazaar *were amongst the top magazines overseas to feature silk designs by the Star of Siam in the late 1950s and early 1960s.*

Today, several internationally renowned Thai designers export ready-made silk evening dresses, suits, blouses and accessories to top boutiques throughout the world. Styles and techniques vary from artist to artist, as do the clientele.

Prabhai Wongseelashote, who designs under the Sue Jay label, specialises in women's fashion. She illustrates the versatility of silk by creating a plethora of garments: business suits, cocktail dresses, casual wear. Khanitha Akaranitikul, of Khanitha Thai Silk, is best known for her glamorous evening gowns. Her work is popular in America; she has an outlet in New York City, the nation's fashion capital. Jean Noel Haxo designs for Bangkok's glitterati, but under the "Star of Siam" label his silk masterpieces have appeared in all the world's capital cities. He believes that for a designer to work successfully with Thai silk, he or she must understand the unique characteristics of the fabric, its strengths and weaknesses and its history.

Somchai Kaewthong (Kai), one of Thailand's brightest young designers, applies personal creativity to the fabric. Hand painting, mixing numerous fabrics, and adding special touches with lace and embroidery, are amongst his trademarks which result in truly innovative designs. Kai prefers Thai silk to the silk of other countries because of the dexterity of its texture and weave. The rich material may be transformed into a variety of forms,

from high-fashion dresses to raincoats, with equal success. Though his techniques are revolutionary, his designs maintain an Oriental style, which he feels matches the fabric.

For fashion conscious individuals who appreciate the texture of delicate old silk, boasting a uniquely artistic pattern, nothing can compare to an exclusive item of apparel made from a length of antique fabric. Designer Napajaree Suanduenchai of Prayer Textile Gallery specialises in combining lengths of old Thai silk with contemporary plain silk to make splendid jackets and dresses. She frequently lines jackets made from other fabrics with silk to confer a natural thickness and warmth. Such jackets are often reversible and can be worn on either side. Napajaree finds that most of her clients are concerned with the artistic function of textiles for clothing and appreciate garments that are traditional, original and not mass produced. She has an enthusiastic following not only in Bangkok, but also in Germany, where she has another outlet located in Berlin.

Tirapan Vanarat, known as "Pom" within fashion circles, is especially fond of *mudmee* silk, which he considers a real work of art. One of Thailand's more popular designers, Tirapan believes selecting a pattern appropriate to the outfit is the secret to creating a stunning design. He likes *mudmee* for both day and evening wear, but insists the designs are simple and classic to highlight the personality and beauty of the wearer.

In modern Thailand, at any grand society ball, wedding or formal occasion, the majority of the gowns will be made of gleaming and opulent silk. Whether traditional Thai costumes or designs that follow the latest Parisian trends, this versatile material lends itself exceptionally well to the fashion world. Here lies the miracle of silk: a transition from the humble caterpillar to a shimmering, exquisite fabric favoured by royalty. More than just a pretty material, Thai silk offers a rustling romance that will whisper endearments of originality forever.

As smooth as silk, Thai Airways International hostesses appear graceful and sophisticated in their silk uniforms.

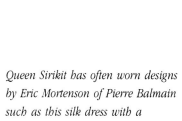

Queen Sirikit has often worn designs by Eric Mortenson of Pierre Balmain such as this silk dress with a mudmee *pattern.*

KAI 93

Kai swirls together rich patterned silks with plain colours to make suits that can be worn anywhere to grace any occasion.

STAR OF SIAM

Evening dresses made of silk are both classical and stunning.

JIM THOMPSON   97

For day-time, night-time or any time, Thai silk garments are comfortable, versatile and practical.

PRAYER

*Strikingly, simply, richly different;
old silks and ethnic cotton can be combined
to provide a high fashion look
that is startlingly original.*

KHANITHA 101

Streamlined and sophisticated, Khanitha turns out clothes that keep their cool as the temperature climbs. Thai silk graces any climate or occasion.

Left: *Use dark coloured silks with strong patterns to create a cosy and intimate ambience (Jim Thompson Thai Silk Co. Ltd.).*

Right: *Silk upholstery on chairs and sofas is not only elegant, but practical as silk is durable and strong.*

Sumptuous Decors Of Silk

It is sophisticated, timeless and beautiful. Long appreciated by connoisseurs of fine living, Thai silk has gained a world wide reputation over the last few decades as an elegant and imaginative interior design fabric.

Long ago, both French and Persian visitors to the city of Ayutthaya remarked on the sumptuous furnishings that included decorative parasols covered in gold cloth and silk and brocade wall hangings. Although used in palaces throughout Thailand for centuries, Thai silk furnishings were, until recently, less common abroad.

After the Second World War, Jim Thompson promoted Thai silk around the world. As a result, in the late 1950s and early 1960s, orders were pouring into his Thai Silk Company. The Savoy Hotel of London chose Thai silk to furnish its suites; it was used in Windsor Castle for refurbishing the Canaletto Room and Woolworth heiress Barbara Hutton selected great quantities of the fabric for her Mexican retreat. Despite the great Chinese silk market on its very doorstep, the Hong Kong Hilton selected Thai silk for its ballroom and its top suites.

Naturally, many of Thailand's most famous hotels, including the Oriental, the Dusit Thani, the Regent and recently the Sukhothai, have chosen to adorn their best suites with the finest Thai silk. As its fame has spread, Thai silk has gained popularity amongst international hotel chains such as the Intercontinental group. They were so impressed by the use of this material in their Bangkok property that they utilised it in Intercontinental hotels in Korea, Oman and Indonesia.

Jackson Lenor Larsen, America's leading textile designer and one of Thai silk's greatest admirers, became the first international designer to use Thai silk purchased from Jim Thompson. Following Thompson's disappearance in 1967, Larsen was invited by the Thai Silk Company to create exclusive designs for silks to be used as furnishing fabrics. Many of his designs have become classics.

Today, well-known Thai interior designers such as Peter Bunnag and Chantika Puranananda use Thai silk in stunningly creative ways. For these top designers, the fabric has become a key medium in the upholstery of both Thai and European style furniture. With the aid of Thai silk, a simple armchair becomes an elegant piece of furniture and the traditional elephant chair takes on a regal appearance.

The late John Rifenberg, a Bangkok-based international architect and interior designer, was renowned for his sumptuous and original soft furnishing schemes in Thai silk. He incorporated the prized material in projects both at home and abroad. Rifenberg's signature was imaginative wall

Right: *This four-poster bed in the Oriental Suite of the Oriental Hotel, Bangkok, designed by the late John Rifenberg, is draped in romantic silk bedhangings with a silk bedspread. The pineapple bed posts add a whimsical touch.*

Page 106-107: *The late John Rifenberg was fond of combining bright silk patterns and plain colours in complementary upholstery colour schemes.*

Above: *Thai silk blends effortlessly with a traditional Thai decor in the living room of Jim Thompson's "house on the klong".*

Right: *In the Oriental Suite at the Oriental Hotel, brightly coloured silk cushions and chairs are shown to advantage against a plain background.*

panelling in Thai silk. In particular, he praised the fabric's suitability for all types of upholstery.

Leading companies specialising in Thai silk home-furnishing fabrics include the Mulberry Company Limited, Design Thai and Thai Silk Company, all of whom export overseas and have agents in major capital cities. Their heavy Thai silks are much admired for a wide range of imaginative colours that vary from plain, classical neutrals to glorious plaids, smart stripes, creative checks, bright abstracts, soft florals and original Thai pattern.

Thai silk owes part of its attraction as an interior design fabric to the fact that plain or neutral colours which would appear dull in other fabrics look stunning when woven in silk because of the rich texture and high quality of the weave.

In addition to beauty, practicality is a large part of the appeal of Thai silk as a decorative fabric. It is not only elegant but also hard wearing and durable; if treated with a fabric protector such as Scotchguard and well cared for, six-ply silk can last for up to 15 years. The material is also extremely versatile, and, as proven by great designers, suitable for an infinite range of uses from wall panelling to furniture upholstery to simple pillow cases. Fashion-wise, it holds a timeless quality. While other fabrics go in and out of style, Thai silk remains a classic.

Before using Thai silk for interior decor, the effects of natural and artificial light must be studied. Under different lights, the colours of the same piece of silk can vary greatly; a colour that may appear harsh under a natural light may be soft and rich beneath artificial light. As silk fades under long exposure to natural light, it is important to consider where it will be utilised when planning a silk-based decorative scheme. Unlined silk is fine for bed hangings that can be positioned away from a source of natural light, but if used for curtains, heavy lining must be added.

Thai silk looks wonderful when contrasted with other natural materials. When combined with rich woods such as mahogany, oak or native Thai teak and rosewood, the colours of silk blend perfectly. Alternatively, this versatile fabric can be used to obtain an ethnic look when earthy colours and designs are combined with rattan. Soft colours can create a warm and inviting ambience.

A popular decorating technique, originated by Jim Thompson, is the use of multi-coloured silk pillow cases. Softly patterned cases contrasted against a cotton sofa can appear simple, yet stylish, while bright, solid-coloured cushions can enliven an otherwise dull room. When framed, a patterned silk square brings an elegant dash of cheerful colour to a wall. Silk frills added to table cloths made from another material, a dramatic pelmet above a wall mirror to match upholstery or cushions, or a length of silk entwined around a chandelier chain will all create a dash of chic.

A four-poster bed can feel even more romantic when shrouded in luscious, smooth, silk bed-hangings with a matching bedspread. Plain or pelmeted curtains are grandly exquisite with the fabric slightly creased and richly spilling onto the floor. Thai silk is perfect for achieving understated simplicity and elegance or stylish sophistication.

Thai *mudmee* silk is a light fabric with brilliant colours and patterns that create superb bed-hangings, bedspreads and wall-hangings. The pattern should be displayed flat to avoid losing any of its impact. *Mudmee* is exhibited to its best advantage when combined with plain silk.

Sensuous, cheerful, exciting and dazzlingly different, Thai silk adds originality and style to any interior design scheme.

Right: *Match silk-lined walls with silk upholstery to achieve a harmoniously subtle decor (Jim Thompson Thai Silk Company).*

Pages 112-113: *For a look that is fun but classical, the late John Rifenberg added silk-covered chairs in a variety of bright colours to this informal living room.*

Right: *Soft lights and subtle silks are the theme at the Sukhothai Bangkok.*

Right: *Sweet dreams in silk at the Jim Thompson Suite in the Oriental Hotel.*

Right: *Draw the bed hangings and sleep within a silk cocoon.*

Left: *Thai silk blends beautifully with Oriental antiques.*

Top: *Silk is used to add the finishing touches to the Royal Suite of the Hotel Siam Inter-Continental, Bangkok.*

•

Above: *Twist swathes of silk around bedframes to add an artistic touch, and match with a coloured bedspread and cushions.*

•

Right: *Yellow silk is bright and uplifting in this classical bedroom (The Mulberry Company, Bangkok).*

Left: *A length of red and gold silk cloth dating to the late 19th century. The supplementary weft is of gold-wrapped thread and the design depicts flowers and mangos based on Malaysian motifs. It is believed to have been woven as a* chung kraben *(length of cloth worn below the waist) for a member of the Thai Royal Family. (Prayer Textile Gallery)*

Right: *From the area of southern Isan, the floral and diamond pattern designs interposed with geometric symbols on these fine silks woven in the early 20th century, display a Lao/Khmer influence.*

Collecting Silk Textiles

Silk textiles are a valuable part of Thailand's cultural heritage. Not only do they afford an insight into early weaving and dyeing techniques, but they also tell a great deal about the history, customs and lifestyles of the Thai people. Silk textiles were indicators of economic status through various periods of history. While cotton was for commoners, rich silks that symbolised influence and power were reserved for the nobility.

Old silk textiles are finer than their contemporary counterparts and often boast brilliant colours due to the use of natural dyes. The styles and patterns of old Thai silk items vary according to the area in which they were made. Fabrics found in Central Thailand frequently depicted Indian styles in their designs, while those woven in the northeastern region of Isan have a strong Lao influence.

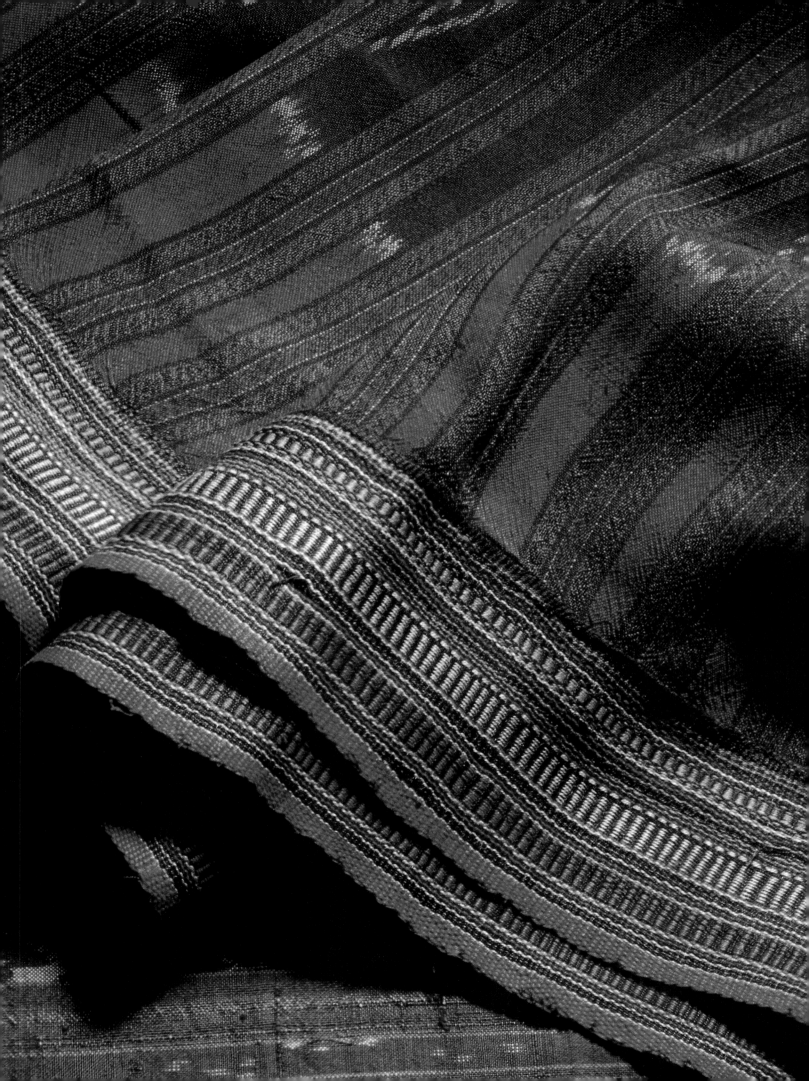

Pages 120-121: *Some forty years old, these silks from the Khon Kaen area have a split pattern divided by stripes. The rich colours owe their origins to natural dyes. (Prayer Textile Gallery)*

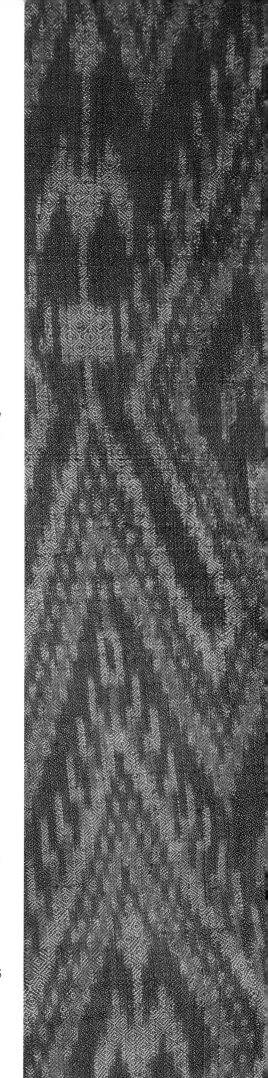

Right: *Woven in Surin and displaying a Khmer influence, the diamond shape and floral pattern utilises traditional colours.*

Because silk textiles were commonly used as articles of clothing, pieces which have survived are often well worn. In addition, the ravages of time along with Thailand's humid, moist climate have made the preservation of silk fabric rather difficult. As a result, antique silks are a rare find, and most pieces in collections around the country are not more than one or two centuries old. In contrast, China boasts silk textiles dating back 4,000 years that have been well preserved in tombs.

Fortunately, for those wishing to enhance their knowledge of textiles or even for those wishing to fulfil their visual appetite, Thailand's various museums house interesting textile exhibits. Bangkok's National Museum has a collection of period silk costumes dating from the Ayutthaya period onwards. Most items were taken from the storerooms of the Royal Household, and are therefore of the highest quality. The collection of silk fabrics includes: a checkered *mudmee* design from the Northeast, satin brocaded with gold and silver threads from India that were made for the Thai market, and a variety woven with Islamic designs created specially for the Royal Household. There are a number of private collections in Thailand, but few are available for regular public viewing.

Outside Bangkok, several provincial museums hold small collections of silk unique to the individual area. Worth visiting are the National Museum

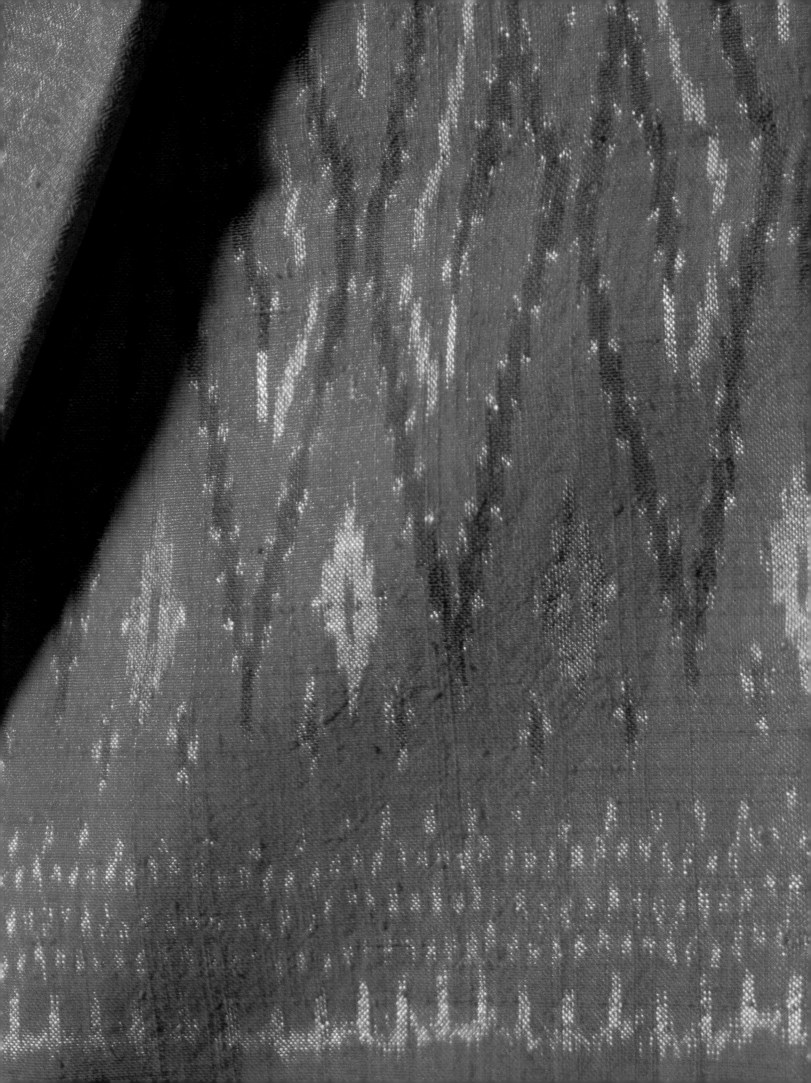

of Chiang Mai where there are fine examples of northern textiles, and the museums of Ubon Ratchathani, Nan, Ratchaburi, Lopburi and Nakhon Si Thammarat. The National Museum in Khon Kaen lies in the heart of silk producing country. Not only does the museum exhibit fine antique silks, but it also includes a collection of weaving implements and artifacts associated with silk production.

Splendid collections of textiles that include some Thai silk artifacts are also exhibited in a number of museums abroad. These include the Anthropology Department of The Smithsonian Institute, the Harvard Peabody Museum of Archaeology and Ethnology, the American Museum of Natural History, the Textile Museum of Washington, the University of Washington and Seattle Museum, and the Australian National Art Gallery.

For textile connoisseurs who are interested in purchasing exquisite old pieces of silk in Thailand, Bangkok has numerous antique shops, several of which specialise in Thai fabrics. The River City Shopping Complex and the Oriental Plaza house the top shops in town, but one should be warned that prized articles are often hidden away in a back room and will only be brought forth and exhibited to customers who display some knowledge of textiles. In the name of adventure, and in search of slightly lower prices, one may choose instead to hunt through the stalls of Chatuchak, Bangkok's weekend market, where rare finds are difficult to come by but always rewarding. Prayer Textile Gallery, on the corner of Phyathai Road and Siam Square, is a favourite venue for local collectors. The owner, Napajaree Suanuenchai, has some fascinating lengths of rare silk and can frequently be persuaded to share her wealth of knowledge on antique silk designs.

Today, fine lengths of antique Thai silk are exceedingly rare. More readily available in Thailand, just as fascinating and considerably less expensive are beautiful pieces from Cambodia and Laos. Usually, only those who are very knowledgeable can distinguish between the silks of these neighbouring countries. When examining a length of antique silk, take it from the confines of the shop and out into the natural light where the colours can be more readily appreciated and repairs detected.

Many Thais continue to incorporate antique textiles into their wardrobe, either as clothing items or as fashion accessories. Lesser antique pieces are sewn into a variety of modern designs, including jackets,

Right: *Check patterns and plaids once common to the Surin area were worn by men as sarongs. Today, such patterns can be found throughout Thailand.*

Pages 124-125: *A pattern that used to be common to Uthai Thani, the red colour owes its vibrancy to natural dyes produced from the skin or shell of the lac insect.*

waistcoats, wall-hangings and scarfs. The largest pieces are found in the form of a lengthy rectangular wrap-around skirt that was worn by both sexes. Woven jackets and wall-hangings or scarf-like lengths of silk that were worn as *sabais* over a blouse are still to be found. They invariably exhibit signs of wear and tear and have often been clumsily repaired.

A length of antique silk may be displayed in various ways, including hanging from a rail or mounted on a frame or as an elegant backdrop for antique artifacts such as Asian silverware.

Collecting old textiles that were painstakingly handwoven heirlooms with fine attention to detail, boasting exquisite patterns created decades ago, can be very rewarding. If correctly cared for, they will endure forever.

PRESERVING ANTIQUE SILK

- Keep the textile out of strong light, away from heat and in a controlled temperature so as to avoid fading and attracting fungus.

- Protect the piece from enviromental hazards such as dust, fumes and cigarette smoke by storing it in a ventilated, dry and enclosed case. It should be rolled but not folded, in acid free tissue paper, cellophane or unbleached muslin.

- If displaying on a wall, use an ultraviolet glass frame or hang for brief periods of time only. Fluorescents can be almost as damaging as sunlight.

- Old silks should be mounted with care. They may be damaged by adhesive chemicals or by potentially rusty nails which cause staining or create holes in weakened fibres. The best means of display is by carefully sewing the item onto a supporting background such as a velcro strip which can be attached to a wooden board. Laying textiles flat avoids strain on the fibres.

- Textiles should be periodically checked for termites or fungi and treated by a professional if damage is detected.

- Cleaning and repair of ancient silk items is a specialised process and must be done by an expert. If the item is strong and in good condition, gently brushing will remove dust particles, and washing the piece in warm water with a suitable non-harsh washing agent will provide a more thorough cleansing. The silk should be dried gently in a flat position.

Right: *Many colour variations of red and purple are used to weave rich and sumptuous patterns.*

Page 130: *The supplementary weft pattern on the red-based silk includes panels of horses and riders, woven in different colours. The other panel depicts the offering tray used in religious ceremonies, dating to the late 1890s.*

A	B
C	D

A: *A simple silk design common to the Ubon Ratchathani area depicts white geometric hooks on a plain background which was used for household decoration.*

•

B: *An early 20th century length of silk uses a silver thread in the supplementary weft, highlighting a dainty diamond and floral design.*

•

C: *These two early 20th century* chung kraben *(garments wrapped around the lower part of the body) were made in Thailand, based on a Khmer design that was common around this time.*

•

D: *A length of red and gold silk dating to the late 19th century. The supplementary weft is of gold-wrapped thread and the design depicts flowers and mangos based on Malaysian motifs. It was thought to have been woven as a* chung kraben *for a member of the Thai Royal Family. (Prayer Textile Gallery)*

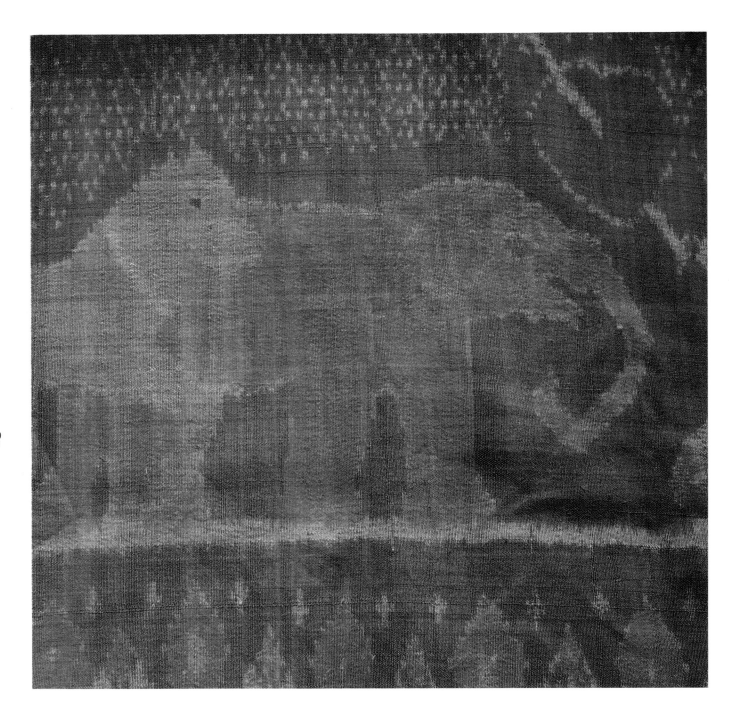

A northeastern design from the Khon Kaen area depicts an elephant and a horse as the central characters. (Prayer Textile Gallery)

A pair of fighting cocks was perhaps derived from the fact that cock-fighting was once a popular sport in the villages of Thailand.
(Prayer Textile Gallery)

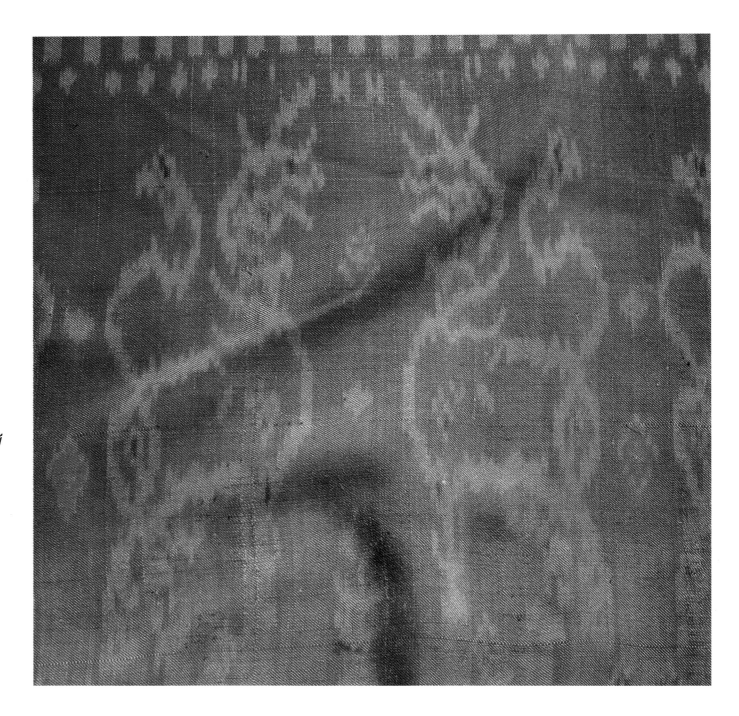

Made around the 1940s, the design depicts the mythological Singh that resembles a tiger.
(Prayer Textile Gallery)

*Originating from Khon Kaen, the weaver incorporated
a pair of parrots into his design.
(Prayer Textile Gallery)*

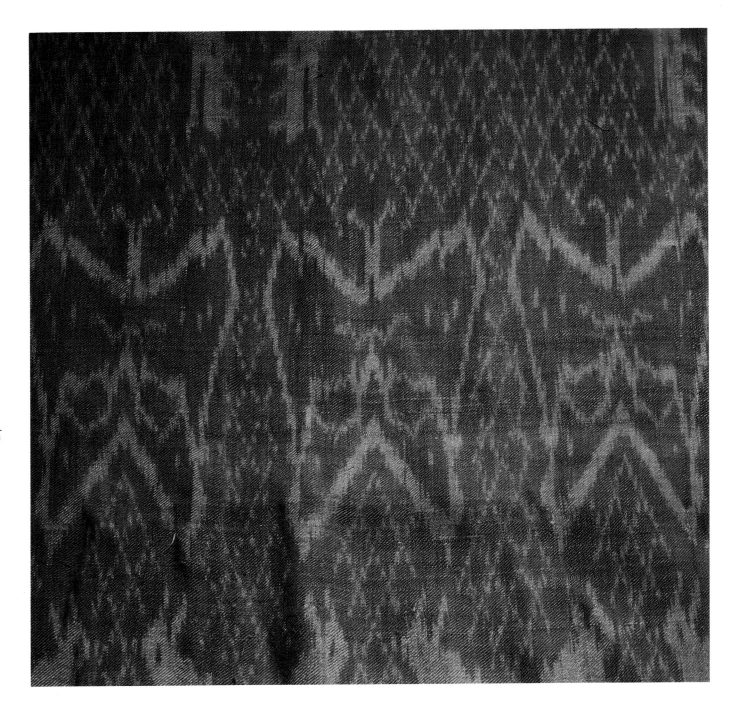

Woven around the early 1950s, nature themes of butterflies and rabbits inspired the weaver of this length of silk from the Northeast.

This silk design was woven at a time when peacocks were a common sight roaming the forests and jungles of Thailand. The joined tails signifies that they are a mating pair.
(Prayer Textile Gallery)

Left: *Four Thai patterns based on Khmer designs that are now rarely found in Thailand. The patterns indicate these are rich silks. (Prayer Textile Gallery)*

Above: *A skirt and fringed shawl made from natural dyes woven in designs common to the northeastern regions of Isan and Khon Kaen.*

Page 140: *A sabai (shoulder cloth) from Isan woven in the early 20th century. The elephants are typical of those found on pieces of silk from Laos and Isan, but are different from the Lanna elephant motifs. (Tilleke & Gibbins collection)*

A	B
C	D

All pieces from Tilleke & Gibbins collection

A: A sabai (shoulder cloth) from Kalasin in Thailand woven in the early 20th century. About 2 metres long, many elements are combined in the weft brocade, including geometrics, floral motifs, rows of spirit figures and snakes.

•

B: A ceremonial cloth from Lanna, woven in the early 20th century. A silk base cloth with supplementary weft weave, it was used to cover a dowry bowl in an engagement ceremony or as a head covering in ordination ceremonies. The central key design is flanked by rows of elephants.

•

C: A ceremonial cloth from Laos woven in the early 20th century. Exceptionally wide and probably part of a much longer piece, this panel may have been used as a ceremonial hanging inside a temple.

•

D: A head cloth from Laos woven in the late 19th century. Two large Khosa Singhs as protection against evil spirits, bordered by birds with a long serpent dominating the centre. Figures are subtly hidden within the pattern as the colour is used to obscure rather than delineate.

The World Of Thai Silk

PROPERTIES OF SILK

Silk contains two properties which make it superior to any other animal fibre: triangular fibres which reflect light like prisms and layers of protein that lend it a smooth sheen. Silk fibre is lustrous, smooth, supple, lightweight, elastic and strong. When viewed longtitudinally, a filament of de-gummed silk appears smooth, lustrous and translucent, whereas raw silk still coated in gum looks bumpy and irregular.

A coarse texture with uneven, slightly knotty threads distinguishes Thai silk from the silk of other countries. It is this quality that makes it perfectly suited to hand-loom weaving.

A silk fibre is highly elastic and resilient. It will stretch by 10 or 20 per cent over its length without damage and will revert to its original shape upon release. Capable of absorbing up to 30 per cent of its own weight in moisture, it can still feel dry. However, a wet fibre must be handled gently, as it loses 20 per cent of its strength. Warm and pleasantly smooth to touch, silk easily absorbs perspiration and is one of the most comfortable fabrics to wear next to the skin. These factors combine to provide a cool-in-summer and warm-in-winter property. Silk's durability enables it to resist mildews, mould and rot that attack fibres. One of the lightest of natural fibres that resists wrinkling, silk when folded takes up very little room.

USES OF SILK

Silk's durability, elasticity and versatility have inspired myriad uses that stretch far beyond clothing.

Since it resists rot, both the Chinese and Egyptians used it in ancient times for wrapping bodies in preparation for burial. The same attribute renders it practical for use in closing sutures in surgery and even for realigning teeth. Being both strong and light, silk is an ideal fabric for making parachutes, as it folds up compactly. Following the Second World War when materials were scarce, enterprising Europeans turned silk parachutes into underwear and other clothing items. When used as a powder bag for a high-calibre gun, silk burns completely leaving no residue.

Silk is a protein fibre and does not conduct heat. This makes it an extremely effective insulator, and it is used for insulating electric wires. As a quilting liner in ski suits, silk is warm to wear because it prevents body heat from dissipating. When used for light-weight clothing such as stockings or lingerie, the fine yarn permits air to pass through the material. Silk's elasticity makes it a perfect material for racing bicycle tyres; it creates smooth traction and endures well. Silk is also used in the production of carpets, astronaut's clothing, sewing thread, fishing lines and typewriter ribbons. Nail wraps, which preserve the condition of varnished finger-nails, are often made from silk. Macerated silk is used in face powder to enhance smoothness, and pupa oil from the silk moth is added to face creams and anti-ageing lotions. In ancient China, silk was used for paper-making.

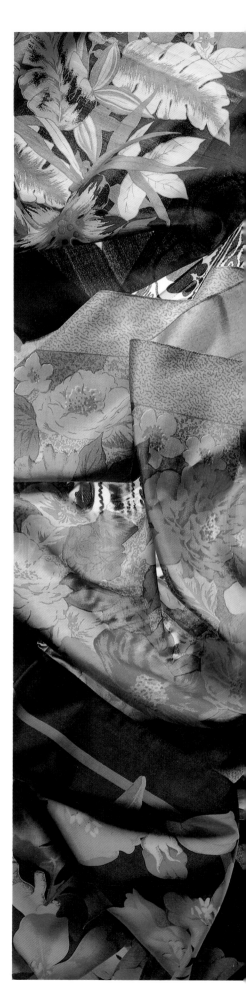

Silk is also a wonderful medium for the visual arts. Painting on silk is an art that has been practised for tens of centuries. Its power of absorbency provides the artist with better control of the colours. This art inevitably led to the modern-day craft of silk screen printing, which was developed in the early 1900s.

SILK BY-PRODUCTS

In Thailand, waste by-products from sericulture can be of considerable value to villagers. A good source of protein, silkworms may be cooked and eaten by humans or used as animal feed. Timber from the mulberry tree is ideal for furniture-making. Dead pupae are used in manufacturing soap and cosmetics, and empty cocoons are decorated and sold as handicraft items.

THE EXPORT MARKET

One of the most expensive fabrics in existence, raw silk is being produced in increasing quantities world-wide: from 55,000 tons in 1980 to 70,000 tons in 1991. Production of silk is dominated by China; Thailand is the world's ninth largest silk producer. Unable to meet a demand for silk thread, in 1991 the Kingdom imported 1,200 tons of fine silk yarns from overseas and produced an estimated 1,000 tons domestically.

Thai silk fabrics and silk products have large foreign and domestic markets. Sixty per cent of the silk is sold locally, worth around US$59 million in 1991. There are over 300 silk exporters, and exports of hand-woven silk and silk products are worth around US$32 million annually. Demand for Thai silk goods overseas is increasing rapidly. The leading importers of Thai silk products are the USA, Japan, Great Britain, Belgium, Australia, Singapore, Switzerland and Germany. The most popular items are neckties, scarfs, ready-made clothes, cushion covers, handicrafts and silk fabric for home furnishing.

The production of Thai silk is a labour-intensive industry that employs hundreds of thousands of people throughout Thailand for the tasks of growing mulberry, raising silkworms, reeling, weaving, dyeing, printing and selling.

THE THAI SILK ASSOCIATION

The Thai Silk Association has over 330 members. They define Thai Silk as: "A pure silk made in Thailand, including Thai silk with raised designs which are woven with any sort of material." This definition excludes fabrics using artificial fibres as well as some materials being made abroad and labelled as Thai silk, both of which are perceived as damaging to the reputation of the genuine product. The numerous retail outlets that satisfy the requirements of the association display certificates of recommendation.

For further information, contact:

The Thai Silk Association
c/o Textile Industry Division
Soi Klueynamthai, Rama 4 Road,
Bangkok 10110, Thailand.
Tel: (662) 391-2896

BUYING AND CARING FOR SILK

A list of recommended retailers and exporters may be obtained from the Thai Silk Association. There are over 1,000 retailers selling silk in Thailand. When buying silk, exercise some prudence and be wary when purchasing silk from other than recommended shops, even in silk producing provinces. A fabric that is sold as "Thai silk" may actually be a type of polyester that has been treated to resemble the genuine material. In the weaving process, polyester is sometimes employed on the warp yarn. The inexperienced buyer may have difficulty detecting this in the finished product. Such a fabric is commonly found in contemporary silk products sold at Chatuchak Weekend Market. A simple burn test may be used to determine whether or not the fabric is genuine silk; when lit with a flame, silk burns with the odour of burning hair or feathers, leaving small easily crushable ash beads that are less regular than beads from synthetic fibres. Cellulose fibres and synthetics may flare up or melt rather than burn.

Genuine silk fabric is sold in six different plies. One ply is most suitable for blouses, two and three ply for dresses and suits, four and five ply for mens' suits or upholstery, and six ply for heavy upholstery. Silk is so durable that, if well cared for, it should last between one and two decades on upholstery. As it ages, it will retain its elegance.

To ensure that silk remains in good condition and maintains its original lustre and texture, it should preferably be dry cleaned. Thai silk may also be washed in luke-warm water but only with the mildest soap. It should be rinsed but not wrung dry. When hung up to dry, the silk should be placed in the shade and supported. A spoonful of clear white vinegar added to the final rinse will maintain the original lustre. Silk should be ironed on the inside of the garment just before it is dry or ironed with a damp cloth applied to the outside. Rich, exotic and oriental, a piece of magnificent Thai silk will always be admired for its beauty.

WHERE TO SEE SILK PRODUCTION

Most silk is produced in the northern and northeastern regions of Thailand. Khon Kaen, Korat and Surin are important centres for fine silk including *mudmee*. There are a number of weaving co-operatives and silk worm production units in these regions. Several large weaving concerns surround the old northern capital of Chiang Mai, particularly in the craft villages of San Kamphaeng some 14 kilometres outside the city. In southern Thailand, silk production can still be seen around Nakorn Si Thammarat.

Silk is also woven in Bangkok, and the process can be viewed at Shinawatra Thai Silk Company's weaving shed on Sukhumvit Soi 23, or H.M. Factory for Thai Silk, Sukhumvit Soi 39 (by appointment only; tel: 258-8766/8769) or in the Bangkrua area along the opposite side of the canal to Jim Thompson's Thai House, where local weavers still practice their skills.

Within driving distance of Bangkok, Her Majesty Queen Sirikit's SUPPORT Project at Bang Sai Art and Crafts Centre near Ayutthaya offers demonstrations of weaving and reeling. Thirty kilometres west of Bangkok, the Rose Garden Country Resort provides demonstrations of silk weaving, spinning and dyeing in a very pleasant garden setting.

Thai Silk souvenirs make beautiful and classical gift items that are appropriate for any occasion.

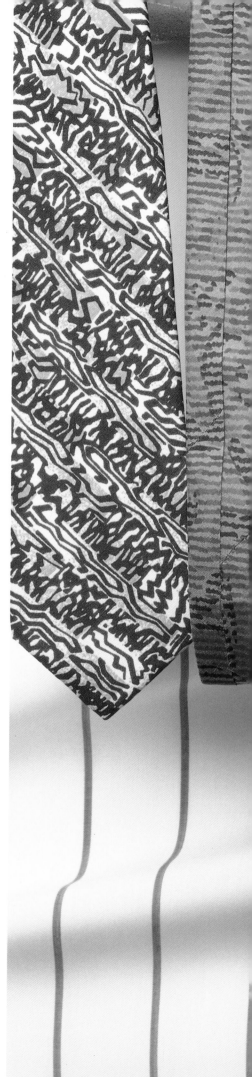

*Silk jackets, shirts and ties are
superb additions to anyone's wardrobe.*

*Pages 154-155: Chatuchak weekend market is a favourite
venue for local residents seeking antique silk bargains.*

GLOSSARY OF SILK TERMS

Ayutthaya	The capital of Siam from 1350 to 1767.
Bivoltine	Silkworm species that can be harvested twice a year.
Bombyx *mori*	The cultivated silkworm that feeds on mulberry leaves and produces the finest white-to-yellow silk.
Brocade	A woven material with a raised pattern, often of interwoven silver or gold.
Chrysalis	The third stage of the silkworm's development between worm and moth.
Cocoon	A small, egg-shaped casing of spun silk, within which the silkworm remains spinning for 23 days.
Fresh cocoon	A cocoon that still contains a live pupa.
Ikat	A term that describes "tie-and-dye" silk weaving where the design is patterned according to the dyeing process.
Chip-na-nang	A style of wearing the *pha nung*. The garment is wrapped around the lower body, gathered in front with both ends folded into pleats and fastened with a belt.
Mudmee	The Thai name for ikat or "tie-and-dye" silk weaving.
Mulberry	The leaves of the mulberry tree are the staple diet of silkworms. Rich in protein, some 220 kilogrammes of mulberry foliage are required to produce one kilogramme of raw silk.
Pha sin	A garment that resembles a skirt which is gathered and folded at the waist before being secured by a belt.
Pha nung	A rectangular piece of material that is tied in the front and folded between the legs *dhoti* style to form a lower garment.
pha nung chung kraben	A style of wearing the *pha nung*. The garment is tied in front, knotted at the waist, then twisted from the top edge to the bottom to form a tail which is folded between the legs and tucked into the waist at the back.

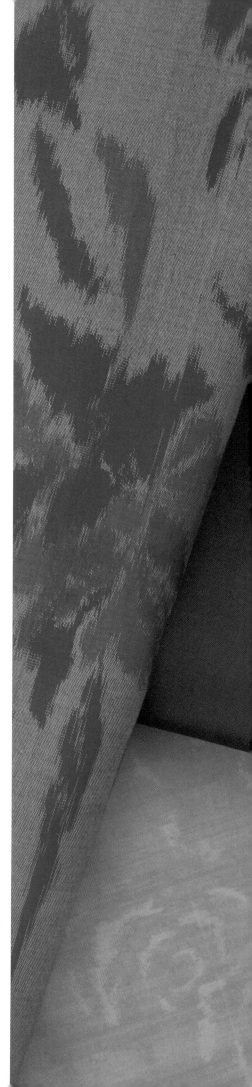

Polyvoltine	Silkworms native to Thailand that can be harvested several times a year.
Pupa	The stage at which the silkworm is enshrouded in the cocoon spinning the filament.
Raw silk	Thread reeled from several cocoons in its natural state.
Reeling	The process of unwinding raw silk filaments from the cocoon.
Sabai	A long piece of material about 30 cms wide that is worn draped across the chest by women.
Saturniidae	The wild silkworm family.
Sericin	A gummy susbstance secreted by the silkworm that holds together the filaments in a cocoon.
Sericulture	The process of raising silk cocoons ready for reeling.
Sin-yok	A cloth woven in gold or silver worn over a Western-style blouse.
Skein	A coil of yarn, usually 114 to 137 centimetres in length.
Slubs	Irregular lumps in silk thread.
Sukhothai	An independent Thai kingdom that escaped Khmer rule and was Thailand's first capital from 1250 to 1438.
Ta-beng-man	A crossed halter top, favoured in Ayuthhaya times.
Throwing	The process where the raw silk is wound from the skein and twisted together to form a yarn.
Tusser	The most common variety of wild silk.
Warp	The threads that pass along the length of the loom.
Weft	The threads that pass from side to side across the warp.

SELECTED BIBLIOGRAPHY

Chongkol Chira
Textiles and Costumes in Thailand
1982, Arts of Asia 12/6.

Chou Ta-Kuan
The Customs of Cambodia
1992, Siam Society, Bangkok

Corbman, Bernard
Textiles, Fibre to Fabric
1985, McGraw-Hill, Singapore

Feltwell, Dr. John
The Story of Silk
1990, Alan Sutton Publishing, U.K.

Finestone, Jeffrey
The Royal Family of Thailand; The Descendents of King Chulalongkorn
1989, Phitsanuloke Publishing Co., Thailand

Fraser-Lu, Sylvia
Handwoven Textiles of Southeast Asia
1988, Oxford University Press, Oxford, Singapore

Geijer, Agnes
A History of Textile Art
1979, Philip Wilson Publishers, London

Gittinger, Mattiebell
Master Dyers to the World
1982, The Textile Museum, Washington D.C.

Gittinger-Lefferts
Textiles and the Tai Experience in Southeast Asia
1992, The Textile Museum, Washington D.C.

Hutchinson, E.W.
Adventurers in Siam in the Seventeenth Century
1985, DD Books, Bangkok

Japan International Co-operation Agency
Silkworm Rearing Techniques in the Tropics
1991, Japan

Kolander, Cheryl
A Silk Worker's Notebook
1985, Interweave Press, Colorado

Laumann, Maryta M.
The Secrets of Excellence in Ancient Chinese Silks
Southern Materials Centre,
1984, Taipei

La Loubere, Simon de
The Kingdom of Siam
1969, Oxford University Press, London, 1980

Maxwell, Robyn
Textiles of Southeast Asia
1990, Oxford University Press

Muqi, Che
The Silk Road, Past and Present
1989, Foreign Languages Press, Beijing

Marks, R. & Robinson A.T.C.
Principles of Weaving
1976, Textile Institute, U.K.

Naenna, Patricia
Costume and Culture
1990, Studio Naenna Co. Ltd.

Narkkong, Apai
Thai Traditional Technology
1987, National Culture Commission

National Museum
An Illustrated Book of Costumes based on Historical and Archaeological Evidence
1968, Fine Arts Department, Bangkok.

Rheinberg, L.
The Romance of Silk
1991, The Textile Institute, U.K.

Syamananda, Rong
A History of Thailand
1990, Chulalongkorn University

Warren, William
The Legendary American
1970, Houghton Mifflin, Boston

Wilson, Kax
A History of Textiles
1979, Westview Press, U.S.A.

ACKNOWLEDGEMENTS

I would like to express my appreciation to Khunying Charungjit Teekara, Lady-in-Waiting to Her Majesty Queen Sirikit of Thailand, who graciously provided photographs of Queen Sirikit and took time to explain the work of the SUPPORT Foundation in detail.

My thanks are due to Mr. Bill Booth of The Thai Silk Company Ltd. for allowing me to visit the company's dyeing and printing operations, and to the following individuals who offered their time, expertise and interest: Khun Prani Obhasanond, Textile Industry Division, Department of Industrial Promotions; Mr. Seigo Yamamoto, Silk City (Thailand) Co. Ltd.; Mr. Peter De Corte, The Textile Institute, Bangkok; Khun Tamrong Sawatwarakul, The Thai Printer & Finishers Co. Ltd.; Mr. Claus Ilg, Regional Adviser on Export Promotions, ESCAP; Khun Duen-Anong Wangwiwatana, Silco Garments Co. Ltd.

I would like to thank Mr. David Lyman of Tilleke & Gibbins, who allowed me to feature some items from his collection of antique textiles and to Ms. Karen Chungyampin, Curator of Tilleke & Gibbins textile collection. Ms. Chungyampin not only shared her knowledge of old textiles but also loaned me research material from her private collection. I am grateful to Khun Napajaree Suanduenchai of Prayer Textile Gallery, who allowed me access to her private collection of rare textiles and agreed to them being featured in this book.

The late Mr. John Rifenberg of Rifenberg & Associates, Bangkok, was one of Thailand's leading architects and interior designers. He was always happy to spend time discussing the use of interior design fabrics and in particular the versatility of Thai silk. Khun Rachnida Hirunpruk, managing director of The Mulberry Co. Ltd., was very generous in sharing her time and knowledge of the use of Thai silk as an interior design fabric and the production processes.

Mrs. Vera Cykman, founder of Star of Siam, related many of her personal memories of Mr. Jim Thompson and the early days of the Thai silk industry. It was fascinating to hear a first-hand account of how Thailand's export market developed from someone who was so actively involved in the industry and contributed to promoting it overseas.

I would like to extend my thanks to other specialists in their own particular fields that covered various aspects of Thai silk, including Thailand's premier fashion designer Khun Kai, French designer Jean Noel Haxo, formerly of Star of Siam, and Mrs. Lea Dingjan-Laarakker for offering some of her weaving knowledge.

Finally I would like to offer my heartfelt thanks to Jane Purananda and her daughter Catherine for their excellent editorial advice and continuous encouragement as the project progressed and to everyone who generously gave assistance and support.

Jennifer Sharples, Bangkok 1993